Using your Maths Question Sets in
Other subjects available online

WHAT'S IN THE BOOK?

Thousands of maths questions

Everything in the book is online at www.lbq.org

www.lbq.org

Use the book to help
find, plan, prepare, rehearse, personalise
AND... learn how you can use lbq.org in
your classroom for...

Whole Class Teaching
Teach mode turns any question into a slide - ideal for whiteboard or touchscreen display

Ad hoc Questioning
Instantly create and send questions to your class

Self Paced Tasks
Set Questions Sets for your class to work through at their own pace.

Teach

If you've got a whiteboard or touch screen you can turn any question into a whole class teaching resource

Wherever you see the 'Teach' symbol you're one click away from turning a question or Question Set into a teaching resource.

TEACH ICON

Each question can be an ideal teaching point

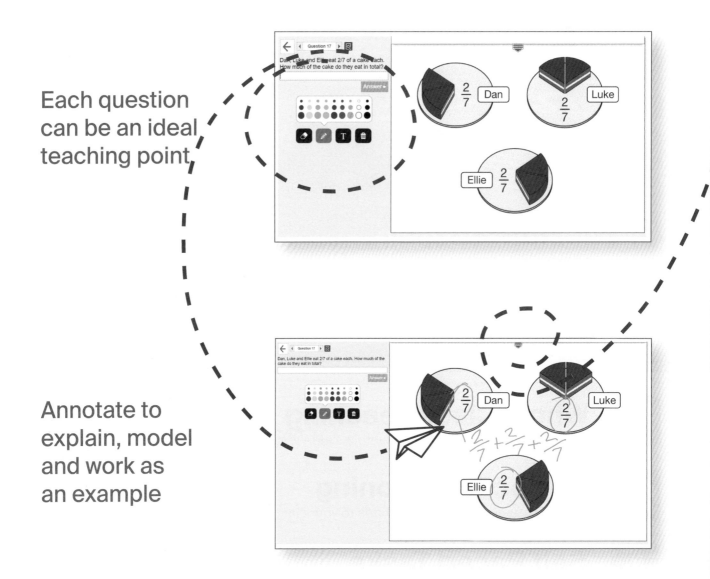

Annotate to explain, model and work as an example

Use the pull down pad to construct your own questions

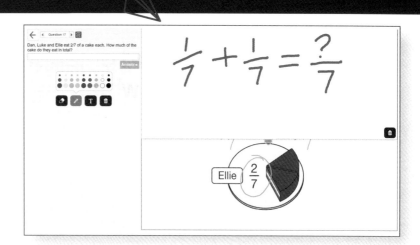

Work through multiple questions like a slide show

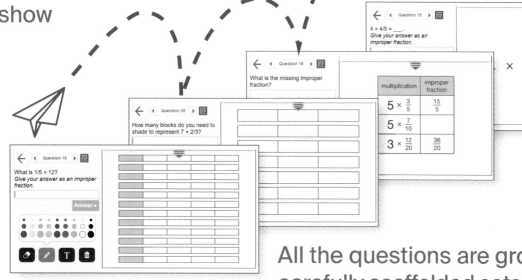

All the questions are grouped into carefully scaffolded sets and provide great support for progression from simple practice to mastery

TEACH MODE IS **FREE** TO ALL REGISTERED USERS

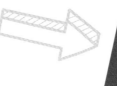

USE THE BOOK TO HELP FIND, ORGANISE AND REHEARSE QUESTIONS TO INCLUDE IN YOUR LESSONS

Ad hoc Questioning*

Have your pupils got tablets, laptops, chromebooks, PCs...? lbq.org makes it fast, easy and super-productive to engage your whole class.

THE AD HOC QUESTION ICON

(always available top right on lbq.org)
is your gateway to asking
questions: at anytime...
            of everyone...
            about anything...

Are your class struggling with a challenging question?

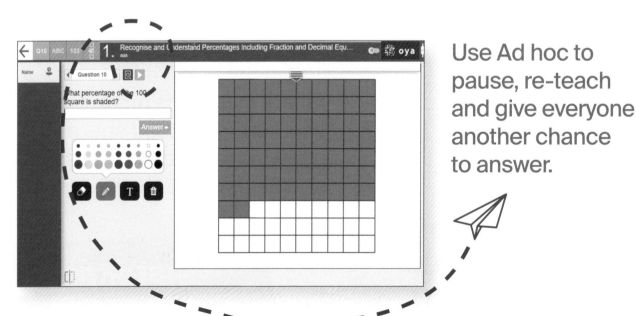

Use Ad hoc to pause, re-teach and give everyone another chance to answer.

*AD HOC QUESTIONING IS **FREE** FOR 60 DAYS!

Want to build on an existing lbq.org question?

Annotate to explain, model, modify and extend...

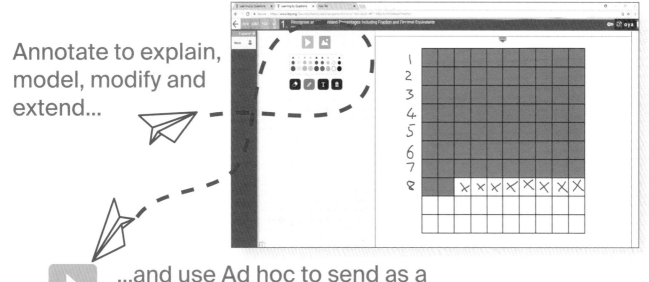

 ...and use Ad hoc to send as a new question to your class.

Or just make your own questions on the fly?

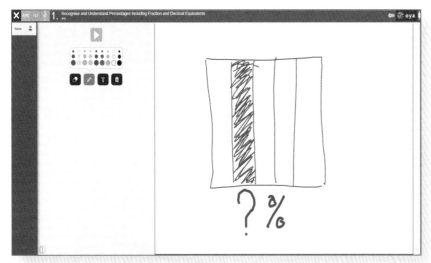

Use our teach tools to create the right question at the right time.

Write a question, draw a question or just ask a question...

... and forget 'hands up'. With Ad hoc questioning everyone answers, every question, every time!

USE THE BOOK TO CHOOSE, ORGANISE AND REHEARSE QUESTIONS TO TEACH IN YOUR CLASSES

Self Paced Tasks*

Lbq.org is built on tens of 1000's of questions - questions grouped into carefully scaffolded sets to provide structured support for learning and to help pin point problems.

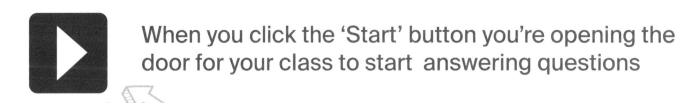

When you click the 'Start' button you're opening the door for your class to start answering questions

START BUTTON

Pupils connect with a simple code and start receiving questions straight away.

Enter your code

6 l d

using LBQ Tasks

App Store | Google play | Get it from Microsoft

*Subscription required for self-paced and ad hoc tasks after initial trial. Teach mode remains free.

Register FREE at lbq.org

And when your pupils hit a challenging question...

... you'll know about it.

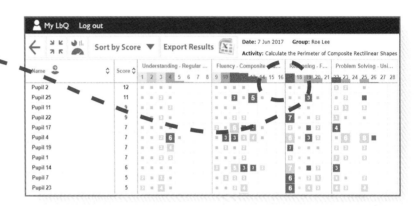

So drill down to see every answer - be right on top of every misconception.

Then pause, intervene, explore, explain, model with the 'Teach' features.

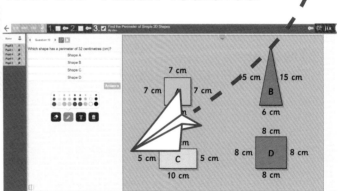

And try the question again with ad hoc questioning.

Everything in this book is online at www.lbq.org

USE THE BOOK TO PICK THE RIGHT TASKS FOR YOUR CLASS AND BE READY FOR INTERVENTION OPPORTUNITIES.

Other titles in the series

Learning by Questions

PRIMARY KS2	SECONDARY KS3
Maths Year 3 Mathematics Primary Question Sets Year 4 Mathematics Primary Question Sets Year 5 Mathematics Primary Question Sets Year 6 Mathematics Primary Question Sets **Science** Years 3&4 Science Lower KS2 Question Sets* Years 5&6 Science Upper KS2 Question Sets* **English** Years 3&4 English Lower KS2 Question Sets* Years 5&6 English Upper KS2 Question Sets*	**Maths** Year 7 Mathematics Primary Question Sets* Year 8 Mathematics Primary Question Sets* Year 9 Mathematics Primary Question Sets* **Biology** KS3 Biology Question Sets* **Chemistry** KS3 Chemistry Question Sets* **Physics** KS3 Physics Question Sets* **English** KS3 English Question Sets*
US Math Grades 4&5 Mathematics Question Sets**	**US Math** Grades 6&7 Mathematics Question Sets**

* Available January 2019
** Available summer 2019

See www.lbq.org/books for title availability

Understanding a question set

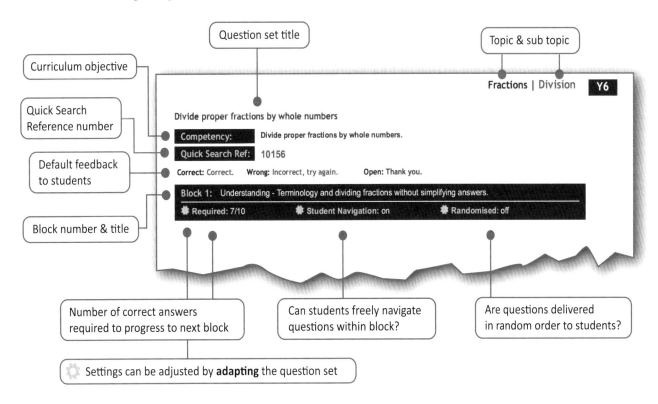

Question set title

Topic & sub topic

Fractions | Division   **Y6**

Curriculum objective

Quick Search Reference number

Default feedback to students

Block number & title

Divide proper fractions by whole numbers

Competency: Divide proper fractions by whole numbers.

Quick Search Ref: 10156

Correct: Correct.   **Wrong:** Incorrect, try again.   **Open:** Thank you.

Block 1:   Understanding - Terminology and dividing fractions without simplifying answers.
✴ Required: 7/10   ✴ Student Navigation: on   ✴ Randomised: off

Number of correct answers required to progress to next block

Can students freely navigate questions within block?

Are questions delivered in random order to students?

⚙ Settings can be adjusted by **adapting** the question set

Understanding a question

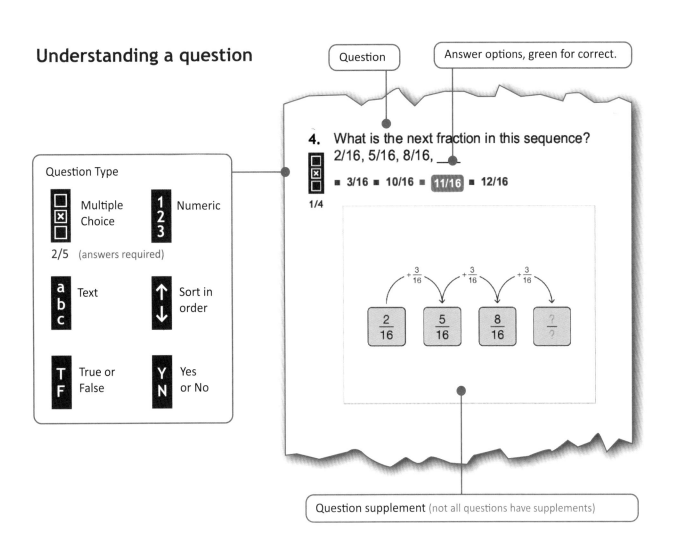

Question

Answer options, green for correct.

Question Type

☐☒☐ Multiple Choice

123 Numeric

2/5 (answers required)

abc Text

↑↓ Sort in order

TF True or False

YN Yes or No

4. What is the next fraction in this sequence?
2/16, 5/16, 8/16, ___

■ 3/16 ■ 10/16 ■ 11/16 ■ 12/16

1/4

Question supplement (not all questions have supplements)

Finding a Question set from this book on the LbQ Platform

The **year group, topic** and **sub topic** classifications used in this book relate directly to those used on the LbQ platform. The fastest way to find a specific question set on lbq.org is via the **Quick Search Reference Number** (e.g. 10654). To further refine a search select a distinctive **keyword** from the question set title or competency.

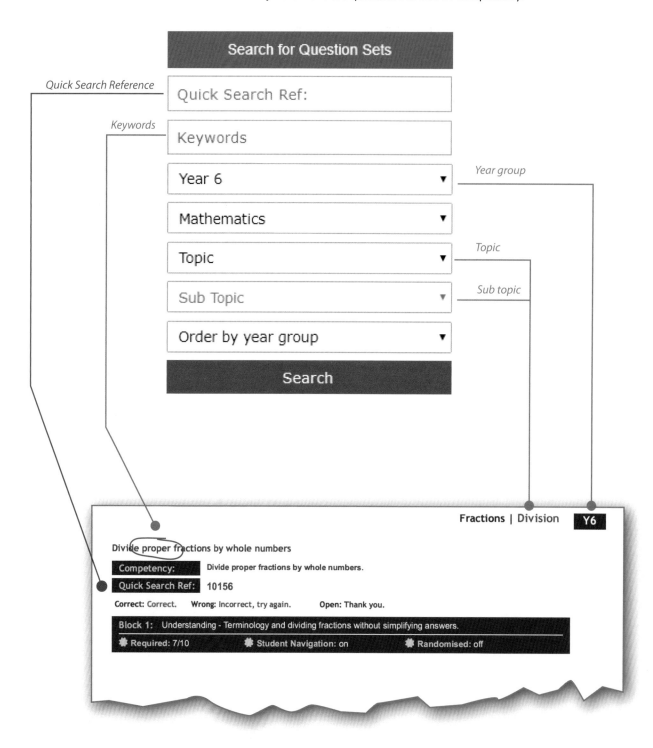

Note: The question sets detailed in this book are correct at time of compilation (July 2018) and correspond directly to the question sets as published on www.lbq.org.

Owing to the nature of www.lbq.org we will from time to time extend, update or modify the question sets published there, which will give rise to discrepancies between this book and the online resources.

Topic Directory **Y8**

Mathematics

Y8

Addition and Subtraction

Mental Calculation

Written Methods Addition

Written Methods Subtraction

Revise four operations using mental methods

Competency: Use the four operations, including formal written methods, applied to integers, decimals, proper and improper fractions and mixed numbers, all both positive and negative.

Quick Search Ref: 10282

Correct: Correct.   Wrong: Incorrect, try again.   Open: Thank you.

Level 1: Understanding - Mental methods for all four operations for positive and negative integers and decimals including multiplying and dividing by positive powers of 10.

✿ **Required:** 7/10   ✿ **Student Navigation:** on   ✿ **Randomised:** off

---

1. Add together 967 and 610.

a b c   ▪ 1,577   ▪ 1577

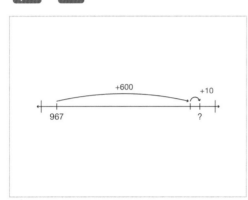

2. What is 59 less than 533?

1 2 3   ▪ 474

3. What is 97 × 5?

1 2 3   ▪ 485

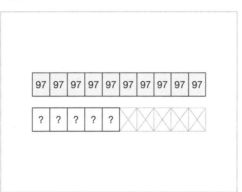

4. Calculate 872 ÷ 4.

1 2 3   ▪ 218

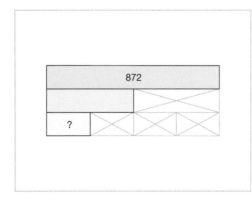

5. Complete the following calculation:

26 - -45 = ___

1 2 3   ▪ 71

6. Find the product of -8 and 6.

1 2 3   ▪ -48

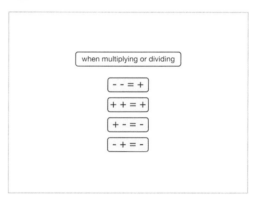

7. Multiply 4.756 by 1,000.

a b c   ▪ 4,756   ▪ 4756

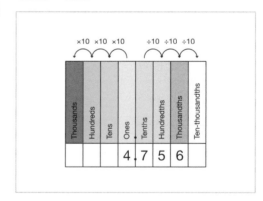

8. Add together 6.5 and 6.6.

1 2 3   ▪ 13.1

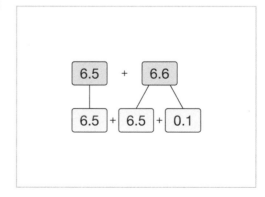

Level 1: cont.

9. What is 7.2 × 11?

$\frac{1}{2}\frac{}{3}$ ▪ 79.2

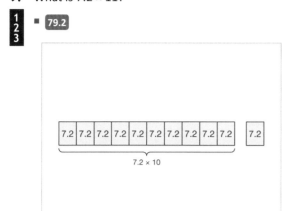

10. Calculate 4,502 + 1,310.

$\frac{a}{b}\frac{}{c}$ ▪ 5,812 ▪ 5812

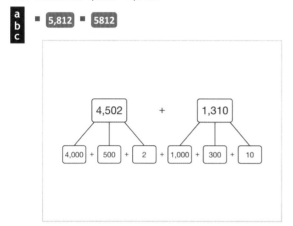

Level 2: Fluency - Mental methods for all four operation including BIDMAS, multiplying and dividing by negative powers of 10 and questions in context.

✿ **Required:** 8/10   ✿ **Student Navigation:** on
✿ **Randomised:** off

11. Use partitioning to work out 520 + 692.

$\frac{a}{b}\frac{}{c}$ ▪ 1,212 ▪ 1212

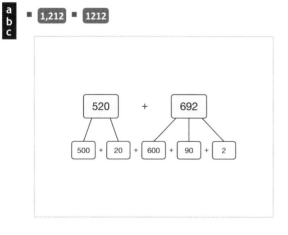

12. What is the product of 32 × 9?

$\frac{1}{2}\frac{}{3}$ ▪ 288

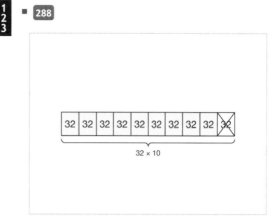

13. ___ - 109 = 823

$\frac{1}{2}\frac{}{3}$ ▪ 932

14. 45,064 ÷ 524 = 86, so which of the following calculations are correct?
There are three correct answers.

3/6 ▪ 86 ÷ 45,064 = 524 ▪ 86 x 524 = 45,064
▪ 5.24 x 8.6 = 4.5064 ▪ 524 ÷ 45,064 = 86
▪ 5.24 x 86 = 450.64 ▪ 524 x 86 = 45,064

15. Work out the answer to 3.057 × 0.01.

$\frac{a}{b}\frac{}{c}$ ▪ 0.03057

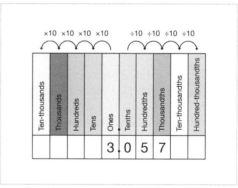

16. Estimate the answer to 1,562 × 83 by rounding to 1 significant figure.

$\frac{a}{b}\frac{}{c}$ ▪ 160000 ▪ 160,000

17. What is 1,059.2 ÷ 0.001?

$\frac{a}{b}\frac{}{c}$ ▪ 1,059,200 ▪ 1059200

Level 2: *cont.*

18. Mr Smith's bank account is overdrawn by £37. If he makes a deposit of £250, what will his new balance be?
Include the £ sign in your answer.

a
b
c

■ £213

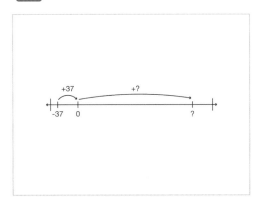

19. Work out the answer to $4 + 7^2 \times 3 - 5$.

1
2
3

■ 146

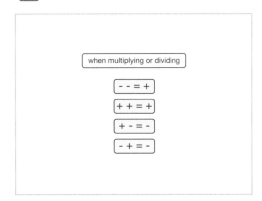

when multiplying or dividing

- - = +
+ + = +
+ - = -
- + = -

20. Sarah wants to buy 2 bags of peanuts and 4 packs of chocolate from the shop. Estimate the amount of money she should take with her in *pounds (£)*.
Include the £ sign in your answer.

a
b
c

■ £12

Level 3: Reasoning - Revise four operations using mental methods.

✱ **Required:** 5/5   ✱ **Student Navigation:** on
✱ **Randomised:** off

21. Fill in the blank:
$240 \times 3 =$ ___ $\times 9$

1
2
3

■ 80

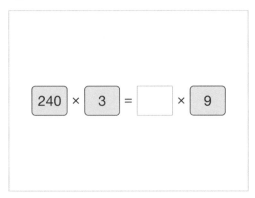

22. Two whole numbers multiply together to give an answer of 600. Neither of the numbers contains a zero. What are the two numbers?

☐
☒
☐

2/5   ■ 24 ■ 15 ■ 48 ■ 25 ■ 12

23. Samuel says 6 + 4 x 9 = 90. Is he correct? Explain your answer.

a
b
c

24. Identify the calculation that is best calculated mentally.

☐
☒
☐

1/4   ■ 475 + 386 ■ 597 + 184 ■ 973 + 168 ■ 834 + 728

25. To divide a number by 4, you half the number 2 times because 4 = 2 × 2. How many times do you half a number by to divide it by 16?

1
2
3

■ 4

Mathematics

Y8

Multiplication and Division

Mental Division

Mental Multiplication

Written Methods Multiplication

Written Methods Division

Revise Four Operations Using Formal Written Methods

Competency: Use the four operations, including formal written methods, applied to integers, decimals, proper and improper fractions and mixed numbers, all both positive and negative.

Quick Search Ref: 10199

Correct: Correct.   Wrong: Incorrect, try again.   Open: Thank you.

Level 1: Understanding - Use formal written methods for the four operations with integers and decimals.

✿ Required: 7/10   ✿ Student Navigation: on   ✿ Randomised: off

1. Add together 46.357 and 1.56.

a b c   ▪ 47.917

2. What is 1.537 less than 2.048?

a b c   ▪ 0.511

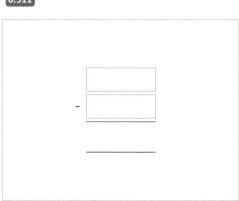

3. Calculate 482 × 234.

a b c   ▪ 112,788 ▪ 112788

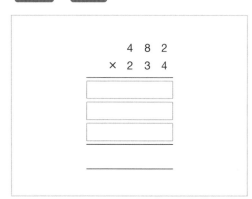

4. What is 966 ÷ 12?

a b c   ▪ 80.5

5. Find the product of 910.1 and 2.5.

a b c   ▪ 2,275.25 ▪ 2275.25

6. Multiply 1.54 by 0.26.

a b c   ▪ 0.4004

7. Divide 869.96 by 14.

a b c   ▪ 62.14

8. Find the difference between 814.32 and 14.97.

a b c   ▪ 799.35

9. Work out the answer to 6.12 × 8.1.

a b c   ▪ 49.572

Level 1: cont.

10. Calculate 7,575.6 ÷ 24.

a b c ▪ 315.65

Level 2: Fluency - Use the order of operations, answer multi-step questions and questions in context.

✺ **Required:** 7/10   ✺ **Student Navigation:** on
✺ **Randomised:** off

11. Perform the following calculation using the column method.
1 2 3 9 - 6.254 = ___.

▪ 2.746

12. Calculate 475.1 + 3.654 + 0.91.

1 2 3 ▪ 479.664

13. 26 + 384 × 14 = ___

a b c ▪ 5,402 ▪ 5402

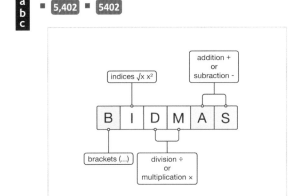

14. What is the missing number?
1 2 3 ___ - 5.28 = 10.9.

▪ 16.18

15. Calculate 37 × 4³ ÷ 2.

a b c ▪ 1184 ▪ 1,184

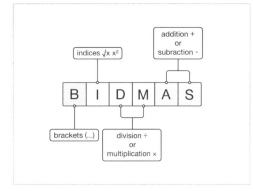

16. Billy has 8 bags of sweets. He eats 2.3 bags himself and his sister eats 1.4 bags. How many bags of sweets does Billy have left?

1 2 3 ▪ 4.3

17. If each table in the canteen seats 8 people, how many tables are needed for 230 people?

a b c ▪ 29

18. It takes Carol 1 minute to swim one length of a swimming pool. If the length of the swimming pool is 48 metres (m), how long does it take her to swim 1 metre in *seconds (s)*?
Include the units in your answer.

a b c ▪ 1.25 seconds ▪ 1.25 s

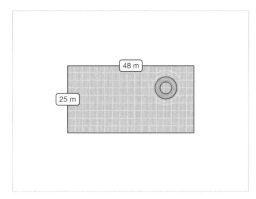

19. The temperature in London is 27°C. The temperature in New York is 0.7 times the temperature in London. What is the temperature in New York in °C?

1 2 3 ▪ 18.9

20. How many 350 gram (g) jars are needed to store 5 kilograms (kg) of marmalade?

1 2 3 ▪ 15

Level 3: Reasoning - Reasoning using formal written methods.

✱ **Required:** 4/4   ✱ **Student Navigation:** on
✱ **Randomised:** off

21. Work out the answer to $24 \times (8 \div 2)^2$.

a
b   ▪ 384
c

22. What is the fifth term in the sequence?

a
b   ▪ 0.75
c

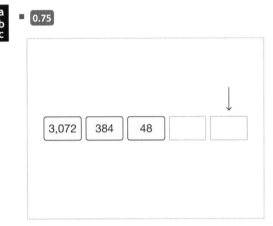

| 3,072 | 384 | 48 | | |

23. Which shop is best value?
Super Saver sells 12 toilet rolls for £1.80.
Top Buys sells 36 toilet rolls for £4.32.
Best Buys sells 48 toilet rolls for £5.28.

1/3

▪ Top Buys ▪ Best Buys ▪ Super Saver

24. Alfie has made a mistake with his calculation. Why did he get the wrong answer and what answer should he have got?

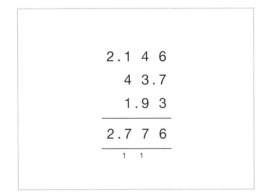

```
  2.1 4 6
    4 3.7
    1.9 3
  ─────────
  2.7 7 6
    1   1
```

Mathematics Y8

Fractions

Multiplication

Division

Fraction Change

Decimals and Percentages

Multiply Fractions and Mixed Number Fractions

Competency: Use the four operations, including formal written methods, applied to integers, decimals, proper and improper fractions and mixed number fractions, all both positive and negative.

Quick Search Ref: 10312

Correct: Correct.   **Wrong:** Incorrect, try again.   **Open:** Thank you.

Level 1: Understanding - Multiplying fractions and mixed number fractions.

✹ Required: 7/10   ✹ Student Navigation: on   ✹ Randomised: off

1. What method is used when multiplying mixed number fractions?

   ▪ Multiply the first numerator by the second denominator and multiply the first denominator by the second numerator.
   ▪ Convert the mixed number fractions to improper fractions and then multiply.
   ▪ Add the whole parts and multiply the fraction parts.
   ▪ Convert the fractions so that they have the same denominator, and then multiply the two numerators.

2. $1\ 1/2 \times 1/4 =$ ___.

   a b c   ▪ 3/8

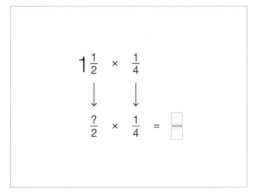

3. Calculate $1\ 2/3 \times 1/7$.

   a b c   ▪ 5/21

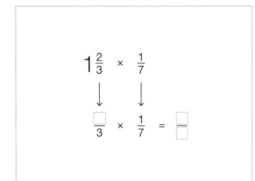

4. What is $1\ 1/10 \times 5/8$?
   Give your answer in its simplest form.

   a b c   ▪ 11/16

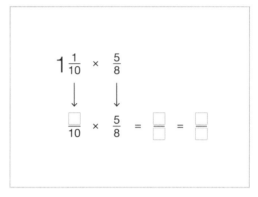

5. What is $2\ 1/3 \times 4/7$?
   Give your answer in its simplest form.

   a b c   ▪ 1 1/3

   $$2\frac{1}{3} \times \frac{4}{7} \rightarrow \frac{\square}{3} \times \frac{4}{7} = \frac{\square}{\square} = \square\frac{\square}{\square}$$

6. Calculate $2/5 \times 1\ 3/7$.
   Give your answer in its simplest form.

   a b c   ▪ 4/7

   $$\frac{2}{5} \times 1\frac{3}{7} \rightarrow \frac{2}{5} \times \frac{\square}{\square} = \frac{\square}{\square} = \frac{\square}{\square}$$

7. 3 1/2 × 4/5 = ___.
a *Give your answer in its simplest form.*
b
c ▪ 2 4/5

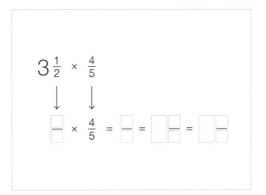

8. 1 2/3 × 4/7 = ___.
a
b ▪ 20/21
c

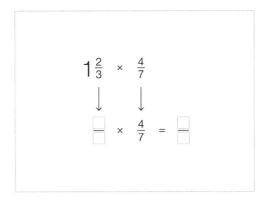

9. Calculate 2 4/5 × 2/5.
a *Give your answer in its simplest form.*
b
c ▪ 1 3/5

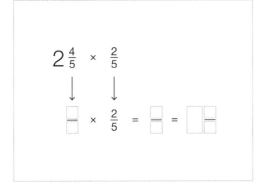

10. What is 3 3/4 × 2/9?
a *Give your answer in its simplest form.*
b
c ▪ 5/6

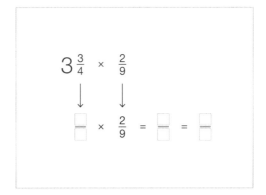

Level 2: Fluency - Multiply fractions and mixed number fractions in context including negative and inverse calculations.

✱ **Required:** 7/10 ✱ **Student Navigation:** on
✱ **Randomised:** off

11. What missing fraction completes the equation?
a 1 4/5 × ___ = 27/45.
b
c ▪ 3/9

12. Curtis's school is 2 1/3 kilometres away. He walks
a 1/8 of the way there and then completes his
b journey by bus. How far does he travel by bus?
c *Give your answer as a mixed number fraction in its simplest form and include the units km (kilometres) in your answer.*

▪ 2 1/24 km ▪ 2 1/24 kilometres

13. What is 1 2/3 × -5/6?

a b c *Give your answer in its simplest form.*

▪ `-1 7/18`

14. A recipe for jam tarts uses 200 grams (g) of jam.
a b c If **2 1/4** pounds are approximately equal to 1 kilogram, how many pounds of jam do you need for the recipe?
Give your answer as a fraction and don't include units with your answer.

▪ `9/20`

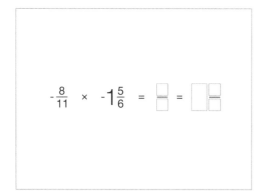

15. Calculate -8/11 × -1 5/6.

a b c *Give your answer in its simplest form.*

▪ `1 1/3`

16. What missing fraction completes the equation?
a b c ___ × 3/5 = 27/35.
Give your answer in its simplest form.

▪ `1 2/7`

17. A smoothie recipe requires 300 millilitres of water.
a b c If 1 litre is approximately equal to **1 3/4** pints, how many pints of water do you need for the recipe?
Give your answer as a fraction in its simplest form and don't include units with your answer.

▪ `21/40`

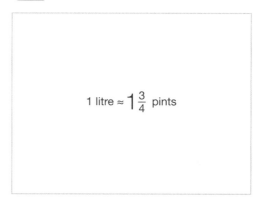

18. Calculate 1 4/7 × 2 1/2.
a b c *Give your answer in its simplest form.*

▪ `3 13/14`

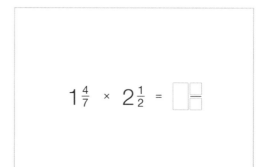

Level 2: *cont.*

19. -7/8 × 2 2/3 = ___.

[a] [b] [c] *Give your answer in its simplest form.*

■ -2 1/3

$$-\frac{7}{8} \times 2\frac{2}{3} = \frac{\square}{\square}$$

20. Calculate 1 2/7 × 3 1/3.

[a] [b] [c] *Give your answer in its simplest form.*

■ 4 2/7

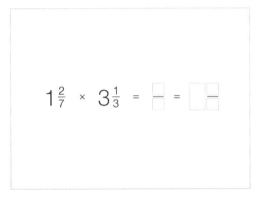

$$1\frac{2}{7} \times 3\frac{1}{3} = \frac{\square}{\square} = \frac{\square}{\square}$$

Level 3: Reasoning - Inverse questions on multiplying fractions and misconceptions.

✿ **Required:** 5/5   ✿ **Student Navigation:** on
✿ **Randomised:** off

21. Multiply the following fractions and arrange them in descending order (largest first).

[↑] [↓]

■ 3 1/2 × 1 3/7   ■ 3 1/3 × 1 1/4   ■ 2 1/2 × 1 1/2
■ 2 1/5 × 1 1/3

22. Select the symbol that completes the following equation:

[□] [⊠] [□]

1 2/7 × 2/3 ___ 2 1/3 × 5/14.

1/3   ■ <   ■ =   ■ >

23. Oliver says that 1 4/9 × 3/5 = 4/5 × 1 3/9.

[a] [b] [c] Is Oliver correct? Explain your answer.

24. When multiplying 4 2/3 and 1/5 Maggie says, "You need to convert the fractions so they have a common denominator". Is Maggie correct? Explain your answer.

[a] [b] [c]

25. Jessica says, "When I multiply two fractions the answer is always a smaller fraction, but when I multiply two mixed number fractions the answer is always greater than the two multipliers". Explain why this is.

[a] [b] [c]

Level 4: Problem Solving - Multi-step problems on multiplying fractions.

✿ **Required:** 5/5   ✿ **Student Navigation:** on
✿ **Randomised:** off

26. In square metres (m²), what is the area of the dotted part of the shape?

[a] [b] [c] *Give your answer as a fraction in its simplest form and don't include the units in your answer.*

■ 3/4

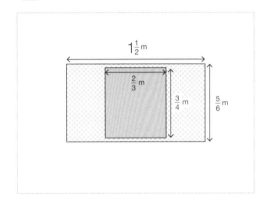

27. In a fraction pyramid, each mixed number fraction is the product of the two mixed number fractions directly below it. What mixed number fraction goes in the top box?

[a] [b] [c]

■ 39 7/12

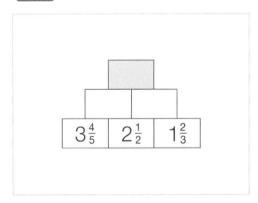

$$3\frac{4}{5} \quad 2\frac{1}{2} \quad 1\frac{2}{3}$$

28. Henry has some marbles and gives 2/3 to Leslie. Leslie now has 11/12 of a bag of marbles. How many bags did Henry start off with?

[a] [b] [c] *Give your answer as a mixed number fraction in its simplest form.*

■ 1 3/8

Level 4: cont.

29. Hannah is building a patio over a quarter of her
a garden. What is the area of her patio?
b *Give your answer as a mixed number fraction in*
c *square metres but don't include the units.*

■ 3 3/20

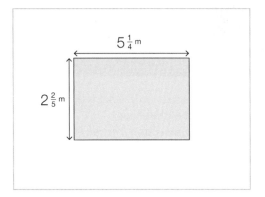

$5\frac{1}{4}$ m

$2\frac{2}{5}$ m

30. What numerator is missing from the first mixed
number fraction?
1
2 2 ?/7 × 3 3/8 = 8 19/28
3

■ 4

Multiply fractions by integers

Competency: Use the four operations, including formal written methods, applied to integers, decimals, proper and improper fractions and mixed numbers, all both positive and negative.

Quick Search Ref: 10013

Correct: Correct.   Wrong: Incorrect, try again.   Open: Thank you.

Level 1: Understanding - Strategy of multiplying proper fractions by positive and negative integers.

✿ **Required:** 7/10   ✿ **Student Navigation:** on   ✿ **Randomised:** off

1. When multiplying a fraction by an integer, what strategy would you use?

1/3
   - ▪ Multiply only the denominator by the integer.
   - ▪ Multiply both the numerator and the denominator by the integer.
   - ▪ Multiply only the numerator by the integer.

2. 4 × 2/9 = ___.

1/3
   ▪ 8/9 ▪ 8/36 ▪ 2/36

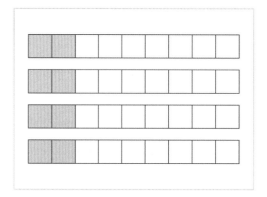

3. Select the 2 statements that are true.
   - ▪ A positive number multiplied by a positive number equals a negative number.
   - ▪ A negative number multiplied by a negative number equals a positive number.
   - ▪ A negative number multiplied by a positive number equals a negative number.
   - ▪ A positive number multiplied a negative number equals a positive number.

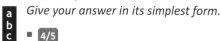

4. Will the answer to -2 × 6/7 be negative or positive?

1/2
   ▪ Negative ▪ Positive

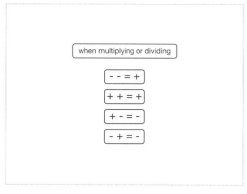

5. What is 2/10 × 4?
   Give your answer in its simplest form.
   ▪ 4/5

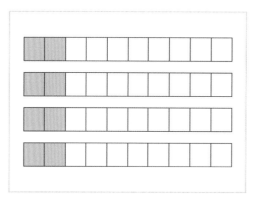

6. Calculate 3/10 × -2.
   Give your answer in its simplest form.
   ▪ -3/5

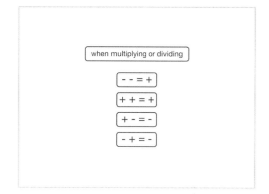

7. -3/11 × -2 = ___.

a
b
c   ▪ 6/11

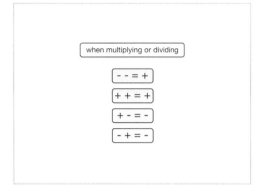

8. Calculate 4/17 of 4.

a
b
c   ▪ 16/17

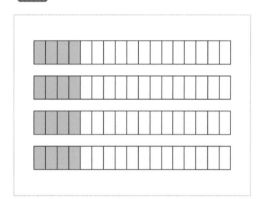

9. In the calculation -5/6 × -4, will the answer be positive or negative?

▪ Positive   ▪ Negative

1/2

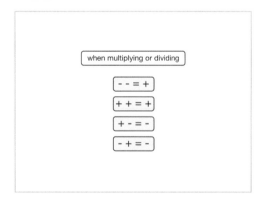

10. -5 × 2/15 = ___.

a
b
c   *Give your answer in its simplest form.*

▪ -2/3

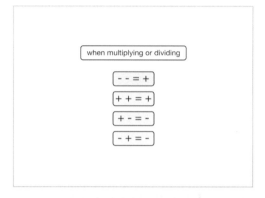

Level 2:   Fluency - Calculations with improper fractions and answers that are equivalent to a whole number.

❄ **Required:** 7/10   ❄ **Student Navigation:** on
❄ **Randomised:** off

11. Layla's cat eats 1/5 of a bag of food each day. What fraction of a bag will she eat in 3 days?

a
b
c   ▪ 3/5

12. What is 83/10 as a mixed number fraction?

a
b
c   ▪ 8 3/10

13. Find 6/7 of 3.

a
b
c   *Give your answer as a mixed number fraction.*

▪ 2 4/7

14. In an art class each student uses 2/5 of a bag of clay. There are 16 students in the class. How much clay will the use altogether?

a
b
c   *Give your answer as a mixed number fraction.*

▪ 6 2/5

15. Calculate 5 × 4/5.

a
b
c   *Give your answer in its simplest form.*

▪ 4

16. Mimi has 9/10 of a punnet of strawberries, Jamie has 4 times this quantity.
How many punnets of strawberries does Jamie have?

a
b
c   *Give your answer as a mixed fraction in its simplest form.*

▪ 3 3/5

17. One batch of onion chutney fills 3/4 of a jar.
If Kayla makes 8 batches, how many jars will she be able to fill with it?

a
b
c   ▪ 6

Level 2: *cont.*

18. What is 3/5 × 4?

a b c *Give your answer as a mixed number fraction.*

■ 2 2/5

19. Calculate 10 × 3/5.

a b c Give your answer in its simplest form.

■ 6

20. 14 × 5/7 = ___.

a b c Give your answer in its simplest form.

■ 10

Level 3: Reasoning - Comparing missing fractions and inverse operations.

✸ **Required:** 4/6   ✸ **Student Navigation:** on
✸ **Randomised:** off

21. 5 × ___ = 2 1/7.

a b c What is the missing fraction?

■ 3/7

22. What calculation is shown by the diagram?

a b c Give your answer in the form **a/b × c**.

■ 5/7 * 4   ■ 5/7 x 4   ■ 4 x 5/7   ■ 4 * 5/7

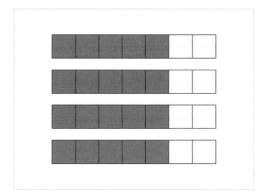

23. Lyla says, "3/4 × -5 is greater than 1/3 × 7". Is she correct? Explain your answer.

a b c

24. Select the symbol that makes the following statement true:

☐ ☒ ☐ 2/7 of 3 ___ 3/7 of 2.

1/3   ■ =   ■ >   ■ <

25. Frankie, Chloe and Ben eat four-fifths of a tin of beans each. What calculation do they need to perform to work out how many tins of beans they ate altogether.

a b c *Give your answer in the form **a/b × c**.*

■ 4/5 * 3   ■ 3 x 4/5   ■ 4/5 x 3   ■ 3 * 4/5

26. Julie says, "5 × 2/7 = 10/35".

a b c Explain the mistakes that she has made.

Level 4: Problem Solving - Multi-step problems multiplying fractions by integers.

✸ **Required:** 6/6   ✸ **Student Navigation:** on
✸ **Randomised:** off

27. Kathryn makes wedding cakes. The table shows how many guests are attending 5 different weddings.

☐ ☒ ☐

3/5 If each guest gets 1/32 of a cake, at which **three** weddings will there be left over cake?

■ Sheffield   ■ Hanson   ■ Caulfield   ■ Hyslop
■ Prestage

wedding	number of guests	amount of cake needed
Sheffield	192	
Hanson	154	
Caulfield	232	
Hyslop	224	
Prestage	98	

28. Isaac is reading a 300 page book. He reads 3/23 of the book each day. After 7 days, how many pages of the book does he have left to read?

a b c *Give your answer as a mixed number fraction.*

■ 26 2/23

29. If **y** × 2/9 is between 28/27 and 4/3, what integer does **y** represent?

1 2 3

■ 5

30. The table shows the ticket prices for a train journey from Manchester to Bolton. How much would it cost to buy one of each ticket in the sale?

a b c *Include the £ sign in your answer.*

■ £16.70

ticket type	price	sale
Adult	£12	$\frac{3}{8}$ off
Child	£7	$\frac{2}{5}$ off
Concession	£10	$\frac{1}{2}$ off

Level 4: *cont.*

31. Pablo puts 4 3/5 gallons of petrol in his car every
a Tuesday.
b On Fridays he puts 5 2/9 gallons of petrol in his
c car.
How much petrol does he put in his car in 6
weeks?
Give your answer as a mixed number fraction.

- 58 14/15

32. In a science class, Drew has a pipette containing

78 millilitres (ml) of water.
He gives 2/7 of it to Nadine.
Nadine already had **double** this amount of water.
How much more water does Nadine have than
Drew, to the nearest whole ml?
Don't include the units in your answer.

- 11

Multiply fractions including mixed number fractions

Competency: Use the four operations, including formal written methods, applied to integers, decimals, proper and improper fractions and mixed number fractions, all both positive and negative.

Quick Search Ref: 10293

Correct: Correct.　　**Wrong:** Incorrect, try again.　　**Open:** Thank you.

Level 1: Understanding - Multiplying fractions and mixed number fractions.

✿ **Required:** 7/10　　　　✿ **Student Navigation:** on　　　　✿ **Randomised:** off

1. What method is used when multiplying two fractions?

1/4

- Multiply the first numerator by the second denominator and multiply the first denominator by the second numerator.
- Add the numerators and multiply the denominators.
- ■ Multiply the two numerators and multiply the two denominators.
- Convert the fractions so that they have the same denominator, and then multiply the two numerators.

2. What is 1/3 × 1/2?

a b c　■ 1/6

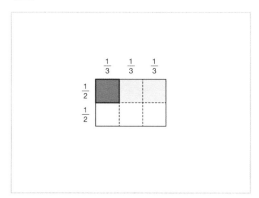

3. Calculate 2/3 × 1/7.

a b c　■ 2/21

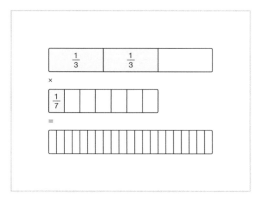

4. 2/3 × 1/5 = ___.

a b c　■ 2/15

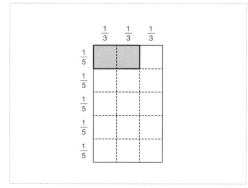

5. What is 1/8 × 4/7?
Give your answer in its simplest form.

a b c　■ 1/14

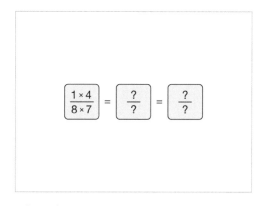

6. 1 1/2 × 1/4 = ___.

a b c　■ 3/8

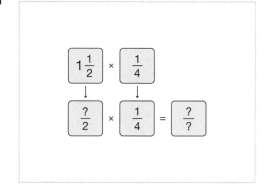

Level 1: cont.

7. Calculate 1 2/3 × 1/7.

 ▪ `5/21`

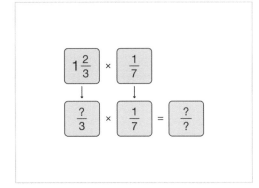

8. What is 7/10 × 5/8?
Give your answer in its simplest form.

▪ `7/16`

9. Calculate 2/5 × 3/7.

 ▪ `6/35`

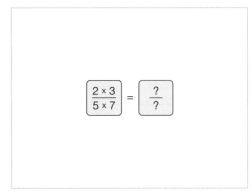

10. 6/7 × 2/9 = ___.
Give your answer in its simplest form.

▪ `4/21`

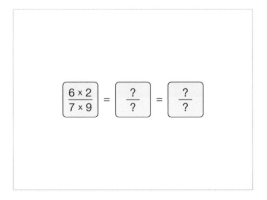

Level 2: Fluency - Multiply fractions in context including negative and inverse calculations.

✱ Required: 7/10   ✱ Student Navigation: on
✱ Randomised: off

11. A pasta recipe for 8 people requires ¾ of a kilogram of meat. Tracey wants to make enough for 3 people. How much meat will she need?
Give your answer as a fraction and include the units kg (kilograms) in your answer.

▪ `9/32 kg` ▪ `9/32 kilograms`

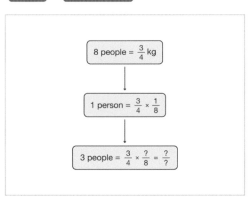

12. What missing fraction completes the equation?
6/11 × ___ = 30/77.

▪ `5/7`

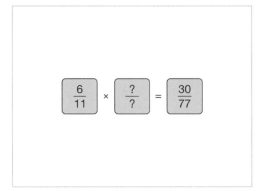

Level 2: cont.

13. Curtis's school is 2 1/3 kilometres away. He walks 1/8 of the way there and then completes his journey by bus. How far does he travel by bus?
Give your answer as a mixed number fraction in its simplest form. and include the units km (kilometres) in your answer.

a b c

■ `2 1/24 km` ■ `2 1/24 kilometres`

14. What is 7/8 × -5/6?

a b c

■ `-35/48`

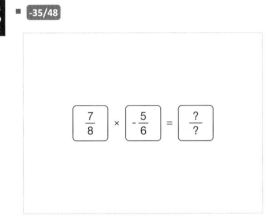

15. What missing fraction completes the equation?
5/7 × ___ = 40/63.

a b c

■ `8/9`

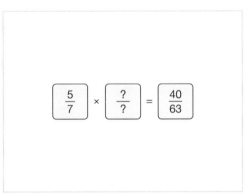

16. A recipe for jam tarts uses 200 grams (g) of jam. If 2 ¼ pounds are approximately equal to 1 kilogram, how many pounds of jam do you need for the recipe?
Give your answer as a fraction and don't include units with your answer.

a b c

■ `9/20`

17. Calculate -8/11 × -5/12.
Give your answer in its simplest form.

a b c

■ `10/33`

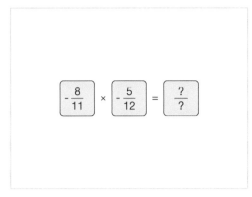

18. What missing fraction completes the equation?
___ × 5/8 = 35/96.

a b c

■ `7/12`

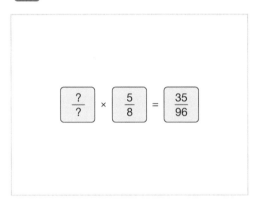

19. A smoothie recipe requires 300 millilitres of water. If 1 litre is approximately equal to 1¾ pints, how many pints of water do you need for the recipe?
Give your answer as a fraction in its simplest form and don't include units with your answer.

a b c

■ `21/40`

Level 2: *cont.*

20. -9/10 × 5/8 = ___.
a *Give your answer in its simplest form.*
b
c ▪ -9/16

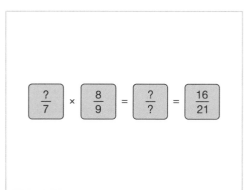

Level 3: Reasoning - Inverse questions on multiplying fractions and misconceptions.

✳ **Required:** 5/5   ✳ **Student Navigation:** on
✳ **Randomised:** off

21. Clare says that 3/11 × 7/8 = 3/8 × 7/11.
a Is Clare correct? Explain your answer.
b
c

22. Laura wrote some notes on multiplying fractions
1 but didn't copy the question correctly.
2 What should the numerator be in the first
3 fraction?
?/7 × 8/9 = 16/21

▪ 6

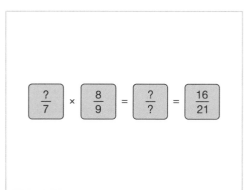

23. Multiply the following fractions and place them in
↑ ascending order (smallest first).
↓ ▪ 1/2 × 11/12 ▪ 3/5 × 4/5 ▪ 3/4 × 2/3 ▪ 2/3 × 7/8

24. To multiply 2/3 and 4/5, Heidi says, "You have
a to convert the fractions so they have a common
b denominator".
c Is Heidi correct? Explain your answer.

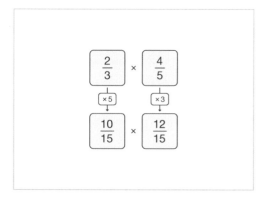

25. What fraction multiplication does the diagram
a represent?
b
c ▪ 4/7 x 5/9 ▪ 4/7 * 5/9 ▪ 5/9 x 4/7 ▪ 5/9 * 4/7

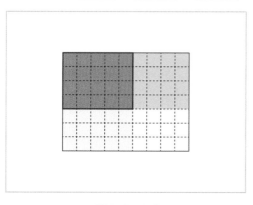

Level 4: Problem Solving - Multi-step problems on multiplying fractions.

✳ **Required:** 5/5   ✳ **Student Navigation:** on
✳ **Randomised:** off

26. In a fraction pyramid, each fraction is the product
a of the two fractions directly below it. What
b fraction goes in the top box?
c *Give your answer in its simplest form.*

▪ 1/27

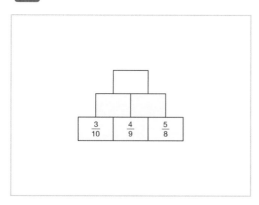

27. Multiply the following fractions and arrange them
↑ in descending order (largest first).
↓ ▪ 3 1/2 × 1 3/7 ▪ 3 1/3 × 1 1/4 ▪ 2 1/2 × 1 1/2
▪ 2 1/5 × 1 1/3

28. The fraction in each square is the product of the
a fractions in the circles on either side.
b What fraction goes in the top circle?
c *Give your answer in its simplest form.*

▪ **2/3**

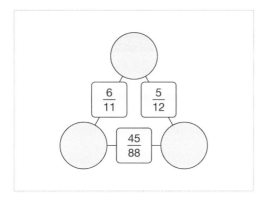

29. In square metres (m²), what is the area of the
a dotted part of the shape?
b *Give your answer as a fraction in its simplest form*
c *and don't include the units in your answer.*

▪ **3/4**

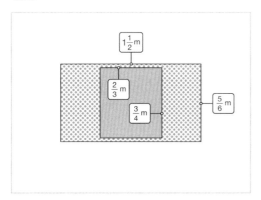

30. Chloe has some peanuts and gives 2/3 of them to
a Bailey.
b Bailey now has 3/5 of a bag of peanuts.
c What fraction of a bag did Chloe start off with?

▪ **9/10**

Multiply Two Fractions

Competency: Use the four operations, including formal written methods, applied to integers, decimals, proper and improper fractions and mixed number fractions, all both positive and negative.

Quick Search Ref: 10311

Correct: Correct.   **Wrong:** Incorrect, try again.   **Open:** Thank you.

Level 1:   Understanding - Multiplying two fractions.

✱ **Required:** 7/10   ✱ **Student Navigation:** on   ✱ **Randomised:** off

1.   What method is used when multiplying two fractions?

1/4

- ■ Multiply the first numerator by the second denominator and multiply the first denominator by the second numerator.
- ■ Add the numerators and multiply the denominators.
- ■ Multiply the two numerators and multiply the two denominators.
- ■ Convert the fractions so that they have the same denominator, and then multiply the two numerators.

2.   What is $1/3 \times 1/2$?

a b c   ■ 1/6

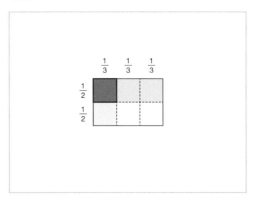

3.   What is $2/3 \times 1/7$?

a b c   ■ 2/21

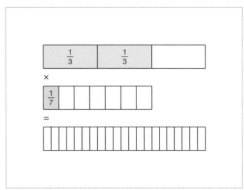

4. $2/3 \times 1/5 =$ ___.

a b c   ■ 2/15

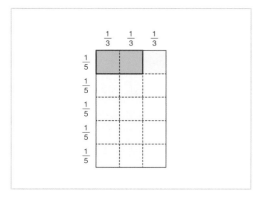

5.   What is $1/8 \times 4/7$?
Give your answer in its simplest form.

a b c   ■ 1/14

$$\frac{1 \times 4}{8 \times 7} = \frac{?}{?} = \frac{?}{?}$$

6.   What is $7/10 \times 5/8$?
Give your answer in its simplest form.

a b c   ■ 7/16

7. Calculate 3/5 × 4/5.

a b c ▪ 12/25

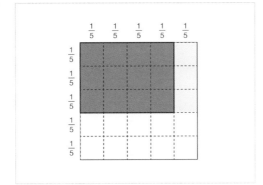

8. 2/7 × 1/4 = ___.

a b c ▪ 2/28

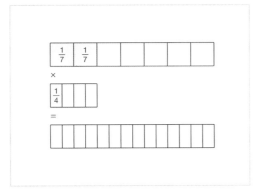

9. Calculate 2/5 × 3/7.

a b c ▪ 6/35

$$\frac{2 \times 3}{5 \times 7} = \frac{?}{?}$$

10. 6/7 × 2/9 = ___.
Give your answer in its simplest form.

a b c ▪ 4/21

$$\frac{6 \times 2}{7 \times 9} = \frac{?}{?} = \frac{?}{?}$$

Level 2: Fluency - Multiply two fractions including negative and inverse calculations.

✱ **Required:** 7/10   ✱ **Student Navigation:** on
✱ **Randomised:** off

11. Amelia usually runs 3/4 of a mile each day. If today she only ran 1/3 of her usual distance, how far has she run?
Give your answer as a fraction of a mile in its simplest form but don't include the units in your answer.

a b c ▪ 1/4

12. What missing fraction completes the equation?
6/11 × ___ = 30/77.

a b c ▪ 5/7

$$\frac{6}{11} \times \frac{?}{?} = \frac{30}{77}$$

13. What is 7/8 × -5/6?

a b c ▪ -35/48

$$\frac{7}{8} \times -\frac{5}{6} = \frac{?}{?}$$

14. If 9/10 of a metre (m) of ribbon wraps four presents, how much ribbon is needed to wrap one present?

a b c

Give your answer as a fraction and include the units m (metres) in your answer.

▪ 9/40 m ▪ 9/40 metre ▪ 9/40 metres

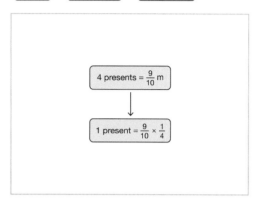

15. What missing fraction completes the equation?

a b c

$5/7 \times$ ___ $= 40/63$.

▪ 8/9

$$\frac{5}{7} \times \frac{?}{?} = \frac{40}{63}$$

16. Calculate $-8/11 \times -5/12$.

a b c

Give your answer in its simplest form.

▪ 10/33

$$-\frac{8}{11} \times -\frac{5}{12} = \frac{?}{?}$$

17. A pasta recipe for **8** people requires **3/4** of a kilogram of meat. Tracey wants to make enough for **3** people. How much meat will she need?

a b c

Give your answer as a fraction and include the units kg (kilograms) in your answer.

▪ 9/32 kilograms ▪ 9/32 kg

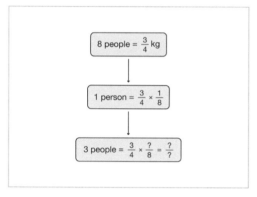

18. What missing fraction completes the equation?

a b c

___ $\times 5/8 = 35/96$.

▪ 7/12

$$\frac{?}{?} \times \frac{5}{8} = \frac{35}{96}$$

19. $-9/10 \times 5/8 =$ ___.

a b c

Give your answer in its simplest form.

▪ -9/16

$$-\frac{9}{10} \times \frac{5}{8} = \frac{?}{?}$$

20. David uses 3/8 of a kilogram of flour to make a
cake. If he halves the recipe how much flour will
he need?
*Give your answer as a fraction and include the
units kg (kilograms) in your answer.*

a b c

▪ 3/16 kg ▪ 3/16 kilograms

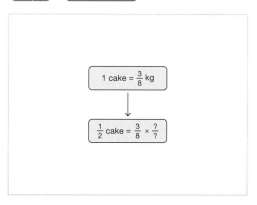

Level 3: Reasoning - Inverse questions on multiplying
fractions and misconceptions.

✱ **Required:** 5/5 ✱ **Student Navigation:** on
✱ **Randomised:** off

21. Clare says that 3/11 × 7/8 = 3/8 × 7/11.
Is Clare correct? Explain your answer.

a b c

22. Laura wrote some notes on multiplying fractions
but didn't copy the question correctly.
What should the numerator be in the first
fraction?
?/7 × 8/9 = 16/21

1 2 3

▪ 6

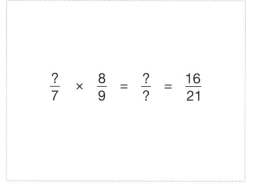

23. Multiply the following fractions and place them in
ascending order (smallest first).

↑ ↓

▪ 1/2 × 11/12 ▪ 3/5 × 4/5 ▪ 3/4 × 2/3 ▪ 2/3 × 7/8

24. To multiply 2/3 and 4/5, Heidi says, "You have
to convert the fractions so they have a common
denominator".
Is Heidi correct? Explain your answer.

a b c

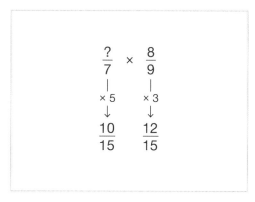

25. What fraction multiplication does the diagram
represent?

a b c

▪ 4/7 * 5/9 ▪ 5/9 x 4/7 ▪ 5/9 * 4/7 ▪ 4/7 x 5/9

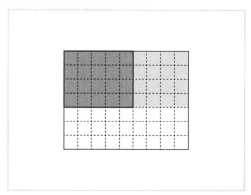

Level 4: Problem Solving - Multi-step problems on
multiplying fractions.

✱ **Required:** 5/5 ✱ **Student Navigation:** on
✱ **Randomised:** off

26. In a fraction pyramid, each fraction is the product
of the two fractions directly below it.
What fraction goes in the top box?
Give your answer in its simplest form.

a b c

▪ 1/27

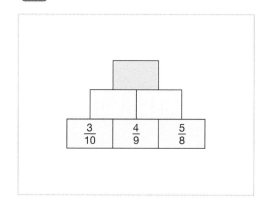

Level 4: *cont.*

27. The fraction in each square is the product of the
a fractions in the circles on either side.
b What fraction goes in the top circle?
c *Give your answer in its simplest form.*

■ 2/3

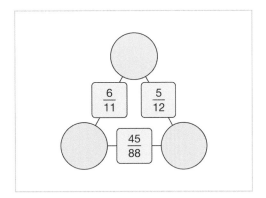

28. What is the area of the triangle in square metres
a (m²)?
b *Give your answer as a fraction in its simplest form*
c *and don't include the units in your answer.*

■ 2/5

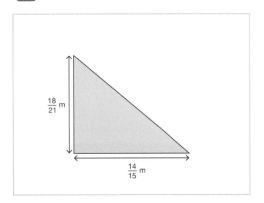

29. Chloe has some peanuts and gives 2/3 of them to
a Bailey. Bailey now has 3/5 of a bag of peanuts.
b What fraction of a bag did Chloe start off with?
c
■ 9/10

30. What is the area of the composite shape in square
a metres (m²)?
b *Give your answer as a fraction in its simplest form*
c *and don't include the units in your answer.*

■ 7/18

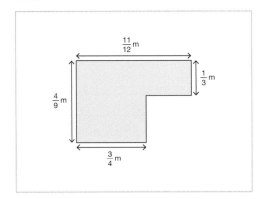

Divide a fraction by a fraction

Competency: Use the four operations, including formal written methods, applied to integers, decimals, proper and improper fractions and mixed numbers, all both positive and negative.

Quick Search Ref: 10232

Correct: Correct.   **Wrong:** Incorrect, try again.   **Open:** Thank you.

Level 1: Understanding - Dividing a fraction by a unit fraction.

❋ **Required:** 7/10   ❋ **Student Navigation:** on   ❋ **Randomised:** off

1. What is a **unit fraction**?

   1/4

   - A whole number and a fraction combined to make one number.
   - A fraction with a denominator of 1.
   - A fraction with a numerator greater than the denominator.
   - A fraction with a numerator of 1.

2. How many eighths are in three-quarters?
   3/4 ÷ 1/8

   - 6

   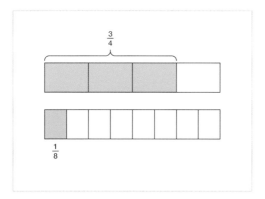

3. How many fifths are in seven-tenths?
   7/10 ÷ 1/5
   Give your answer as a mixed number fraction in its simplest form.

   - 3 1/2

   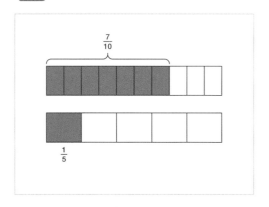

4. What is the method for dividing a fraction by a unit fraction?

   1/3

   - Multiply the numerators and multiply the denominators.
   - Convert the two fractions so that they have common denominators and then add together the fractions.
   - Multiply the fraction by the denominator of the unit fraction.

5. How could you calculate how many thirds are in five-sixths?
   5/6 ÷ 1/3

   - 5/6 ÷ 3 ■ 5/6 x 1/3 ■ 5/6 x 3 ■ 6/5 ÷ 3

6. How many fifths are in seven-eighths?
   7/8 ÷ 1/5
   Give your answer as an improper fraction.

   - 35/8

   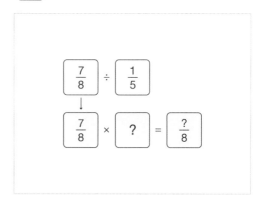

7. How many sixths are in six-sevenths?
   6/7 ÷ 1/6
   Give your answer as an improper fraction.

   - 36/7

   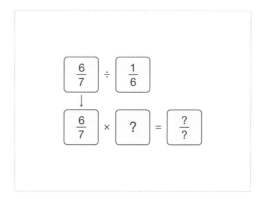

8. How many fifths are in two-thirds?

a
b 2/3 ÷ 1/5
c Give your answer as a mixed number fraction.

▪ 3 1/3

$$\frac{2}{3} \div \frac{1}{5}$$
$$\downarrow$$
$$\frac{2}{3} \times \frac{?}{} = \frac{?}{?} = ?\frac{?}{?}$$

9. How many thirds are in three-quarters?

a
b 3/4 ÷ 1/3
c Give your answer as an improper fraction.

▪ 9/4

$$\frac{3}{4} \div \frac{1}{3}$$
$$\downarrow$$
$$\frac{3}{4} \times \frac{?}{} = \frac{?}{?}$$

10. How many sevenths are in three-fifths?

a
b 3/5 ÷ 1/7
c Give your answer as a mixed number fraction.

▪ 4 1/5

$$\frac{3}{5} \div \frac{1}{7}$$
$$\downarrow$$
$$\frac{3}{5} \times \frac{?}{} = \frac{?}{?} = ?\frac{?}{?}$$

✶ Required: 7/10 ✶ Student Navigation: on
✶ Randomised: off

11. 4 1/2 ÷ 1/4 = 18

a
b Use this information to calculate 4 1/2 ÷ 3/4.
c
▪ 6

$$4\frac{1}{2} \div \frac{1}{4} = 18$$
$$4\frac{1}{2} \div \frac{3}{4} = ?$$

12. 2 1/2 ÷ 1/6 = 15

a
b What is 2 1/2 ÷ 5/6?
c
▪ 3

$$2\frac{1}{2} \div \frac{1}{6} = 15$$
$$2\frac{1}{2} \div \frac{5}{6} = ?$$

13. How could you calculate 4/5 ÷ 3/7?

☐
☒ ▪ 4/5 x 7 ÷ 3 ▪ 4/5 x 3 ÷ 7 ▪ 4/5 ÷ 3 ÷ 7
☐
1/3

14. Calculate 3/8 ÷ 5/7.

a
b ▪ 21/40
c

$$\frac{3}{8} \div \frac{5}{7}$$
$$\downarrow$$
$$\frac{3}{8} \times 7 \div 5 = \frac{?}{?}$$

Level 2: *cont.*

15. One foot is approximately 3/10 of a metre.

a b c Use this information to find how many feet are in 3/4 of a metre.

Give your answer as a mixed number fraction in its simplest form.

Don't include units with your answer.

■ 2 1/2

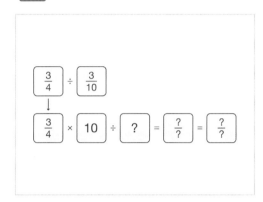

16. What is 2 3/4 ÷ 2/7?

a b c Give your answer as an improper fraction.

■ 77/8

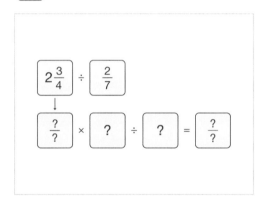

17. What missing fraction makes the following

a b c equation correct?

3/8 ÷ ___ = 33/40.

■ 5/11

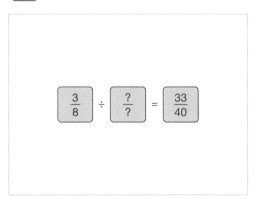

18. Calculate 19/3 ÷ 5/6.

a b c Give your answer as a mixed number fraction in its simplest form.

■ 7 3/5

19. A kilometre is approximately 5/8 of a mile. Use

a b c this information to find how many kilometres are in 4 1/2 miles.

Give your answer as an improper fraction in its simplest form.

Include units in your answer.

■ 36/5 kilometres   ■ 36/5 km

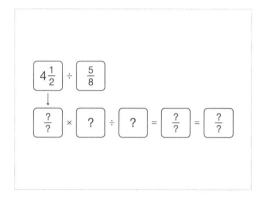

20. What is 4 2/3 ÷ 5/6?

a b c Give your answer as a mixed number fraction in its simplest form.

■ 5 3/5

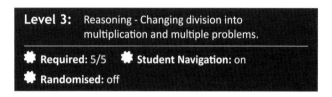

Level 3: Reasoning - Changing division into multiplication and multiple problems.

✱ **Required:** 5/5   ✱ **Student Navigation:** on

✱ **Randomised:** off

21. Jay says, "To divide a fraction by a fraction, you can

a b c change the divide to a multiply and invert the second fraction".

Is Jay correct? Explain your answer.

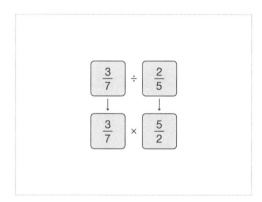

Level 3: *cont.*

22. What missing fraction makes the following
equation true?

5/9 ÷ 7/8 = 5/9 x ___.

■ 8/7   ■ 1 1/7

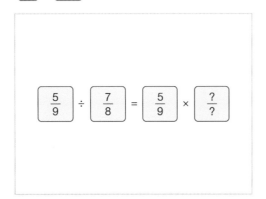

23. Which two calculations give the same answer.

■ 2/9 x 3/5   ■ 2/9 x 5/3   ■ 9/2 ÷ 3/5   ■ 9/2 x 3/5

■ 2/9 ÷ 3/5

2/5

24. Zainub says that 59/63 ÷ 59/63 = 101/103.
Is Zainub correct? Explain your answer.

25. Calculate the following and arrange them in
descending order (largest first).

■ 3/20 ÷ 2/9   ■ 1/2 ÷ 4/5   ■ 3/8 ÷ 5/8   ■ 1/2 ÷ 7/8

Level 4:   Problem Solving - Multi-step questions.

✹ **Required:** 5/5   ✹ **Student Navigation:** on
✹ **Randomised:** off

26. The area of a wall is 12 5/6 square metres (m²).
If the height of the wall is 2 3/4 metres, what is
the length?
Give your answer in metres as a mixed number
fraction.
Include the units in your answer.

■ 4 2/3 m   ■ 4 2/3 metres

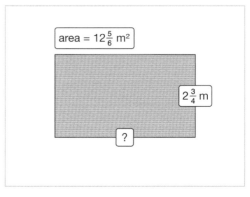

27. What missing fraction makes the following
equation correct?
Give your answer in its simplest form.
___ x 5/12= 3/8

■ 9/10

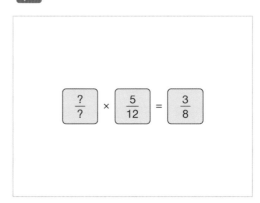

28. Calculate the following and arrange in **ascending**
order (smallest at the top).

■ 11/12 ÷ 2/3   ■ 1 3/4 ÷ 1 1/5   ■ 1 1/3 ÷ 8/9

■ 6/11 ÷ 1/3

29. Lewis and Leyla have identical football sticker
albums. Lewis has filled 2/5 of his album. Leyla has
completed 3/4 of hers.
How many times more stickers does Leyla have
than Lewis?
Give your answer as a mixed number fraction.

■ 1 7/8

Level 4: *cont.*

30. What missing mixed number fraction makes the
following equation correct:

5 3/5 ÷ ___ = 2 2/3

■ 2 1/10

$$5\frac{3}{5} \div \frac{?}{?} = 2\frac{2}{3}$$

Divide an integer by a fraction

Competency: Use the four operations, including formal written methods, applied to integers, decimals, proper and improper fractions and mixed numbers, all both positive and negative.

Quick Search Ref: 10288

Correct: Correct.   Wrong: Incorrect, try again?   Open: Thank you.

Level 1: Understanding - Dividing integers by unit fractions and proper fractions.

✿ Required: 7/10   ✿ Student Navigation: on   ✿ Randomised: off

1. What is a unit fraction?

   1/4
   - A fraction with a numerator greater in value than the denominator.
   - A fraction with a denominator of 1.
   - A fraction with a numerator of 1.
   - A whole number and a fraction combined to make one number.

2. Which calculation would you use to find out how many halves are in 4?

   1/4
   - 1/2 ÷ 4 ▪ 4 x 1/2 ▪ 4 - 1/2 ▪ 4 ÷ 1/2

3. How many halves are in 6?
   6 ÷ 1/2

   - 12

   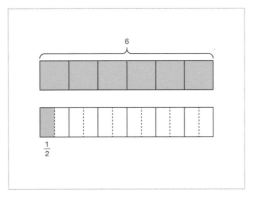

4. What method is used to divide an integer by a unit fraction?

   1/2
   - Multiply the integer by the denominator of the unit fraction.
   - Divide the integer by the denominator of the unit fraction.

5. How many quarters are in 7?
   7 ÷ 1/4

   a b c
   - 28

   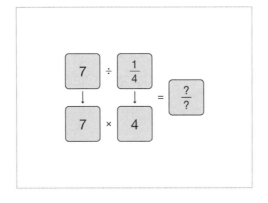

6. What is 4 ÷ 2/3?

   - 6

   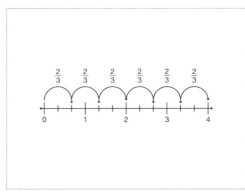

7. Calculate 3 ÷ 3/4.

   a b c
   - 4

   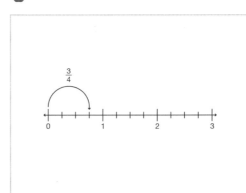

8. Calculate 4 ÷ 2/5.

   a b c
   - 10

   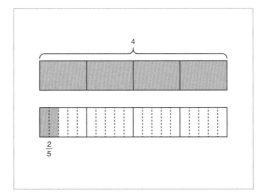

9. How many fifths are in 3?

 3 ÷ 1/5

■ **15**

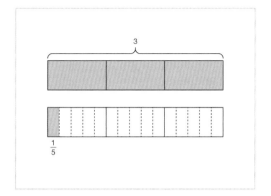

10. What is 6 ÷ 1/3?

 ■ **18**

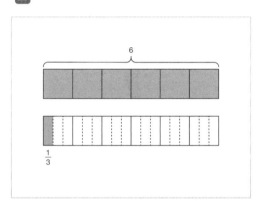

Level 2: Fluency - Dividing integers by proper fractions giving fractional answers.

✿ **Required:** 7/10   ✿ **Student Navigation:** on
✿ **Randomised:** off

11. Use the number line to answer the following:

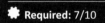 3 ÷ 2/3

Give your answer as a mixed number fraction.

■ **4 1/2**

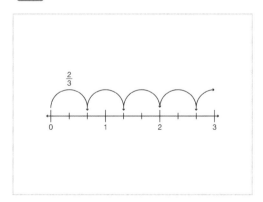

12. Use the diagram to answer the following:

 2 ÷ 3/4

Give your answer as a improper fraction.

■ **8/3**

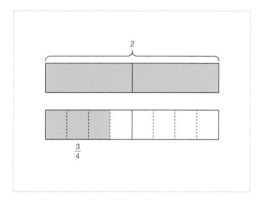

13. A cat eats 2/3 of a tin of cat food per day. How many days will a pack of 8 tins last?

■ **12**

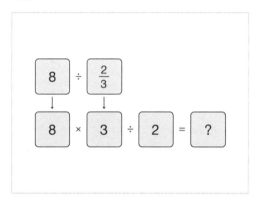

14. What is 7 ÷ 2/3?

Give your answer as an improper fraction.

■ **21/2**

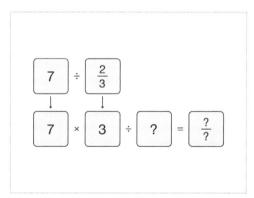

Level 2: *cont.*

15. A cyclist can travel 2/5 of a mile in one minute.
a How long would it take the cyclist to travel 15
b miles?
c Give your answer in minutes as an improper
fraction.
Don't include units in your answer.

▪ **75/2**

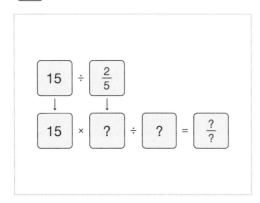

16. What is 3 ÷ 2/5?
a Give your answer as a mixed number fraction.
b
c ▪ **7 1/2**

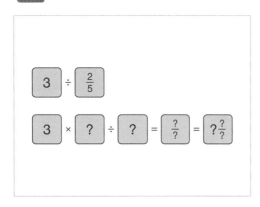

17. A farmer's fence panels are 5/6 of a metre long.
a How many fence panels will she need to enclose a
b square field with a 220 metre perimeter?
c
▪ **264**

18. Calculate 9 ÷ 4/5.
a Write your answer as an improper fraction.
b
c ▪ **45/4**

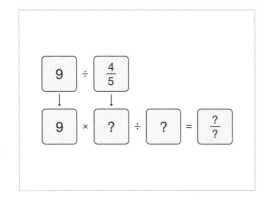

19. A kilometre is approximately 3/5 of one mile.
a Use this information to estimate how many
b kilometres are in 26 miles.
c Give your answer as a mixed number fraction.
Include the units km (kilometres) in your answer.

▪ **43 1/3 km** ▪ **43 1/3 kilometres**

20. Calculate 9 ÷ 6/7.
a Give your answer as a mixed number fraction in its
b simplest form.
c
▪ **10 1/2**

Level 3: Reasoning - Inverse questions and multiple
calculations with reasoning.

✱ **Required:** 5/5   ✱ **Student Navigation:** on
✱ **Randomised:** off

21. What fraction makes the following equation true?
a 9 ÷ ___ = 12 3/5
b
c ▪ **5/7**

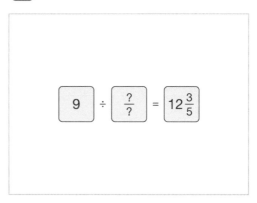

22. William says, "To divide an integer by a fraction, I
a can change the divide to a multiply and invert the
b fraction".
c Is William correct? Explain your answer.

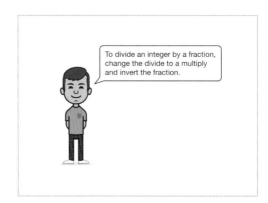

23. Two of the following calculations give the same
a answer.
b What is the answer of the odd one out?
c 8 ÷ 14/15
7 ÷ 7/9
6 ÷ 7/10

▪ **9**

24. Raul says, "If I divide any number in the four times
a table by three-quarters, I will get an integer
b answer".
c Is Raul correct? Explain your answer.

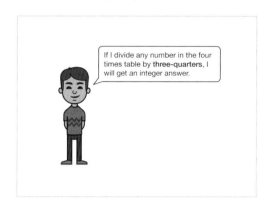

25. Calculate the following and arrange them
↑ in descending order (largest first).
↓
▪ 5 ÷ 12/13 ▪ 3 ÷ 5/9 ▪ 4 ÷ 3/4

Level 4: Problem Solving - Multi-step problems
including questions in context.

✱ **Required:** 5/5   ✱ **Student Navigation:** on
✱ **Randomised:** off

26. Complete the calculations and arrange them in
↑ ascending order (smallest first).
↓
▪ 5 ÷ 2 2/5 ▪ 7 ÷ 3 1/3 ▪ 3 ÷ 1 1/4 ▪ 4 ÷ 1 1/2

27. Darryl paints a wall 7 metres high and 32 metres
a long.
b He buys 10 tins of paint and each tin will cover 23
c 1/3 square metres (m²).
How many tins of paint will Darryl have left over?
Give your answer as a fraction.

▪ 2/5

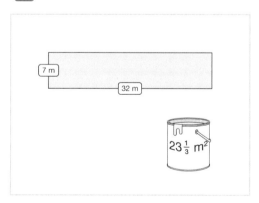

28. What missing mixed number fraction makes the
a following equation true?
b $18 ÷ \_\_\_ = 14\ 2/5$
c

▪ 1 1/4

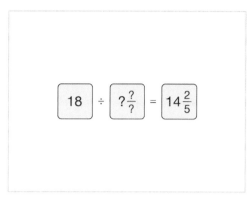

29. A rectangular garden has an area of 30 square
a metres (m²) and a width of 4 1/6 metres.
b What is the length of the garden in centimetres?
c *Include the units cm (centimetres) in your answer.*

▪ 720 cm ▪ 720 centimetres

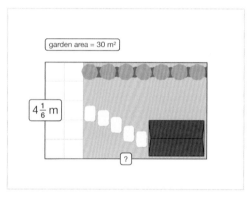

30. Cassie is on a canal barge travelling to her friend's
a boat 18 miles away.
b If she sets off at 11:00 in the morning travelling at
c 3 3/4 miles per hour, what time will arrive with her
friend?
*Give your answer using the 24-hour clock e.g. 1:00
p.m. is given as 13:00.*

▪ 15:48 ▪ 15.48

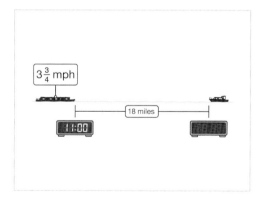

Divide fractions and mixed number fractions by integers

Competency: Use the four operations, including formal written methods, applied to integers, decimals, proper and improper fractions and mixed numbers, all both positive and negative.

Quick Search Ref: 10209

Correct: Correct.   Wrong: Incorrect, try again.   Open: Thank you.

Level 1: Understanding - Dividing a fraction by an integer, including simplifying.

✳ **Required:** 7/10   ✳ **Student Navigation:** on   ✳ **Randomised:** off

1. What is the method for dividing a fraction by an integer?

1/4

- Divide the denominator by the integer; the numerator stays the same.
- Multiply the numerator by the integer; the denominator stays the same.
- Divide the numerator by the integer; the denominator stays the same.
- Multiply the denominator by the integer; the numerator stays the same.

2. What is 1/6 ÷ 2?

1/4

- 2/6 ▪ 1/12 ▪ 2/12 ▪ 1/3

3. When you calculate 2/3 ÷ 5, what will the denominator be?

- 15

4. Use the diagram to calculate 1/3 ÷ 2.

- 1/6

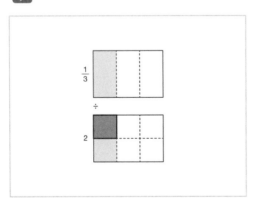

5. What is 1/4 ÷ 2?

- 1/8

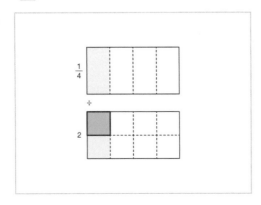

6. Calculate 3/4 ÷ 6.
What is the numerator of your answer in its simplest form?

- 1

7. What is 6/7 ÷ 4?
Give your answer as a fraction in its simplest form.

- 3/14

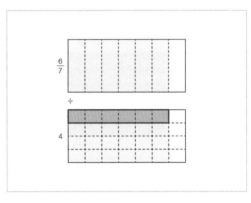

8. Use the diagram to calculate 3/5 ÷ 6.
Give your answer as a fraction in its simplest form.

- 1/10

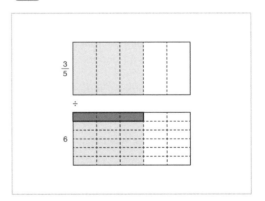

9. Calculate 7/9 ÷ 3.

- 7/27

10. Calculate 10/13 ÷ 15.
Give your answer as a fraction in its simplest form.

- 2/39

Level 2: Fluency - Dividing a fraction by an integer in context, dividing mixed numbers by integers and inverse questions.

✱ **Required:** 7/10   ✱ **Student Navigation:** on
✱ **Randomised:** off

11. What is 1 1/3 ÷ 5?

a b c ▪ 4/15

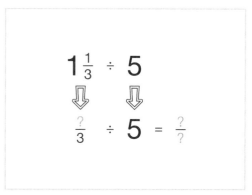

$$1\frac{1}{3} \div 5$$
$$\frac{?}{3} \div 5 = \frac{?}{?}$$

12. What is 1 1/7 ÷ 6?

a b c ▪ 4/21

$$1\frac{1}{7} \div 6$$
$$\frac{?}{7} \div 6 = \frac{?}{?} = \frac{?}{?}$$

13. What is 1 1/4 ÷ 3?

a b c ▪ 5/12

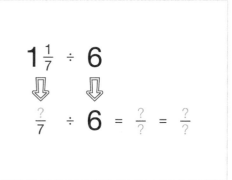

$$1\frac{1}{4} \div 3$$
$$\frac{?}{4} \div 3 = \frac{?}{?}$$

14. A recipe requires 1/4 of a kilogram of self-raising flour to make 12 scones. How much self-raising flour do you need to make one scone?
Give your answer as a fraction and include the units kg (kilogram) in your answer.

▪ 1/48 kg ▪ 1/48 of a kg ▪ 1/48 of a kilogram
▪ 1/48 kilograms ▪ 1/48 kilogram

15. Calculate 1 1/6 ÷ 3.

a b c ▪ 7/18

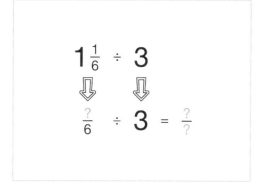

$$1\frac{1}{6} \div 3$$
$$\frac{?}{6} \div 3 = \frac{?}{?}$$

16. Carol completes an 8 1/4 mile walk in three hours. How far does Carol walk in one hour?
Give your answer in miles as a mixed number. Don't include the units in your answer.

a b c

▪ 2 3/4

17. What missing integer makes the following statement correct?
2/5 ÷ ? = 2/35

1 2 3

▪ 7

$$\frac{2}{5} \div \,? = \frac{2}{35}$$

18. What missing fraction makes the following statement correct?
? ÷ 4 = 3/28

a b c

▪ 3/7

$$\frac{?}{?} \div 4 = \frac{3}{28}$$

Level 2: cont.

19. What is 2 1/4 ÷ 5?

a
b
c
    ▪ 9/20

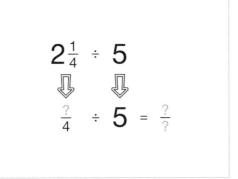

$$2\frac{1}{4} \div 5$$
$$\Downarrow \qquad \Downarrow$$
$$\frac{?}{4} \div 5 = \frac{?}{?}$$

20. What is 3 5/12 ÷ 4?

a
b
c
    ▪ 41/48

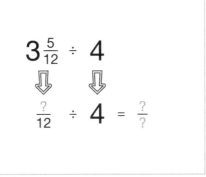

$$3\frac{5}{12} \div 4$$
$$\Downarrow \qquad \Downarrow$$
$$\frac{?}{12} \div 4 = \frac{?}{?}$$

Level 3: Reasoning - Multiple division questions and misconceptions.

✳ **Required:** 5/5   ✳ **Student Navigation:** on
✳ **Randomised:** off

21. What missing integer makes the following equation correct?

1
2
3
    4 1/2 ÷ ? = 3/4

    ▪ 6

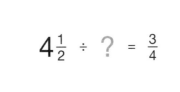

$$4\frac{1}{2} \div \; ? \; = \frac{3}{4}$$

22. To cross the river, calculate the answer to each of the problems and arrange the stepping stones in ascending order (smallest first).

↑
↓

    ▪ 3/10 ÷ 5 ▪ 3/8 ÷ 6 ▪ 2/3 ÷ 8 ▪ 5/16 ÷ 3 ▪ 2/3 ÷ 6

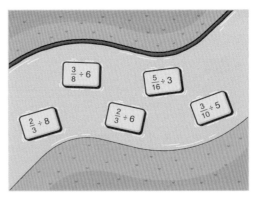

23. Which symbol makes the following statement correct?

☐
☒
☐
    4 5/9 ÷ 8 ____ 5 1/4 ÷ 9

1/3   ▪ < ▪ > ▪ =

24. Robyn says, "I can't divide 5/12 by 3 because 5 is not divisible by 3". Is Robyn correct? Explain your answer.

a
b
c

I can't divide $\frac{5}{12}$ by 3 because 5 is not divisible by 3.

25. Which two of the calculations are incorrect?

☐
☒
☐
    ▪ 3/7 ÷ 5 = 3/35 ▪ 14/15 ÷ 8 = 7/60 ▪ 15/22 ÷ 6 = 5/42
    ▪ 1 7/8 ÷ 12 = 5/32 ▪ 5 1/4 ÷ 12 = 5/12

2/5

Level 4: Problem Solving - Problems involving dividing fractions and mixed numbers by an integer.

✳ **Required:** 5/5   ✳ **Student Navigation:** on
✳ **Randomised:** off

26. Arrange the calculations in ascending order (smallest first).

↑
↓

    ▪ 1 7/8 ÷ 3 ▪ 3 1/5 ÷ 5 ▪ 5 1/3 ÷ 8 ▪ 4 1/2 ÷ 6
    ▪ 2 4/7 ÷ 3

27. A rectangular garden has an area of 115.5 square
a metres (m²). The length of the garden is 14 metres
b (m). What is the width?
c *Give your as a mixed number fraction and include*
the units m (metres) in your answer.

- 8 1/4 metres ▪ 8 1/4 m

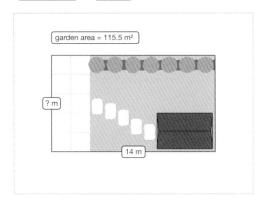

28. Debbie buys a 2 litre bottle of water and drinks
a 350 millilitres (ml). She lets her three friends share
b the remaining water. If they each take an equal
c amount, what fraction of a litre does each friend
receive?
Give your answer in its simplest form. Don't
include the units in your answer.

- 11/20

29. The heights of 3 trees are given in metres. What is
a their mean average height?
b Tree (a): 1 1/2 metres.
c Tree (b): 2 1/4 metres.
Tree (c): 1 2/3 metres.
Give your answer as a mixed number fraction in its
simplest form and include the units m (metres) in
your answer.

- 1 29/36 metres ▪ 1 29/36 m

30. Laura thinks of a fraction and performs the
a following calculations on it:
b +1, ×2, +3, ×4, +5, ×6
c Laura's answer is 182. What was her original
fraction?

- 2/3

Divide Fractions by Integers

Competency: Use the four operations, including formal written methods, applied to integers, decimals, proper and improper fractions and mixed numbers, all both positive and negative.

Quick Search Ref: 10336

Correct: Correct.   **Wrong:** Incorrect, try again.   **Open:** Thank you.

Level 1: Understanding - Dividing a fraction by an integer, including simplifying.

✱ **Required:** 7/10   ✱ **Student Navigation:** on   ✱ **Randomised:** off

1. What is the method for dividing a fraction by an integer?

1/4

- ▪ Divide the denominator by the integer; the numerator stays the same.
- ▪ Multiply the numerator by the integer; the denominator stays the same.
- ▪ Divide the numerator by the integer; the denominator stays the same.
- ▪ Multiply the denominator by the integer; the numerator stays the same.

2. What is 1/6 ÷ 2?

1/4

▪ 2/6 ▪ 1/12 ▪ 2/12 ▪ 1/3

3. When you calculate 2/3 ÷ 5, what will the denominator be?

▪ 15

4. Use the diagram to calculate 1/3 ÷ 2.

abc ▪ 1/6

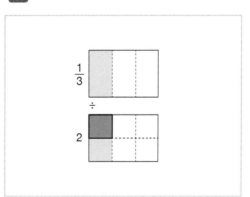

5. What is 1/4 ÷ 2?

abc ▪ 1/8

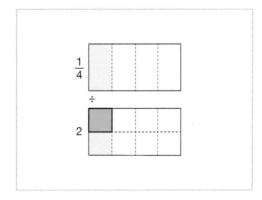

6. Calculate 3/4 ÷ 6.
What is the numerator of your answer in its simplest form?

▪ 1

7. What is 6/7 ÷ 4?
Give your answer as a fraction in its simplest form.

abc ▪ 3/14

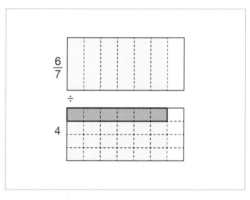

8. Use the diagram to calculate 3/5 ÷ 6.
Give your answer as a fraction in its simplest form.

abc ▪ 1/10

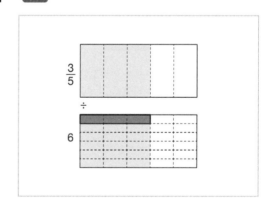

9. Calculate 7/9 ÷ 3.

abc ▪ 7/27

10. Calculate 10/13 ÷ 15.
Give your answer as a fraction in its simplest form.

abc ▪ 2/39

Level 2: Fluency - Dividing a fraction by an integer in context and inverse questions.

✹ **Required: 7/10**   ✹ **Student Navigation: on**
✹ **Randomised: off**

11. What missing integer makes the following statement correct?

[1/2/3] 2/5 ÷ ? = 2/35

▪ **7**

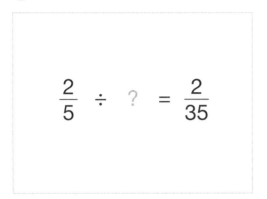

$$\frac{2}{5} \div ? = \frac{2}{35}$$

12. What calculation does the diagram represent?

[☐/☒/☐] ▪ 1/4 ÷ 12 ▪ 1/12 ÷ 4 ▪ **1/4 ÷ 3** ▪ 1/3 ÷ 4

1/4

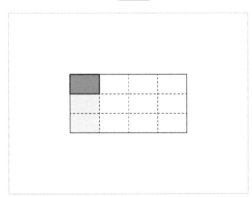

13. A recipe requires 1/4 of a kilogram (kg) of self-raising flour to make 12 scones. How much self-raising flour do you need to make one scone?
Give your answer as a fraction. Don't include the units in your answer.

▪ **1/48**

14. Sally uses 2/3 of a metre (m) of ribbon to wrap eight identical presents. How much ribbon is needed to wrap one present?
Give your answer in its simplest form and include the units m (metres) in your answer.

▪ **1/12 metre** ▪ **1/12 m** ▪ **1/12 metres**

15. Calculate three-eighths divided by five.
Give your answer as a fraction in digits.

▪ **3/40**

16. What missing fraction makes the following statement correct?

[a/b/c] ? ÷ 4 = 3/28

▪ **3/7**

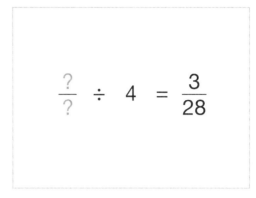

$$\frac{?}{?} \div 4 = \frac{3}{28}$$

17. Select the two options that make the following statement true:
__ ÷ __ = 1/12.

[☐/☒/☐]
2/6   ▪ **2/3** ▪ 3 ▪ **8** ▪ 1/6 ▪ 2/7 ▪ 4

18. Julie completes an 5/6 mile run in three minutes. How far does Julie run in one minute?
Give your answer in miles and don't include the units in your answer.

▪ **5/18**

19. Calculate five-ninths divided by three.
Give your answer as a fraction in digits.

▪ **5/27**

20. What calculation does the diagram represent?

[☐/☒/☐] ▪ 1/5 ÷ 4 ▪ 2/5 ÷ 3 ▪ **2/5 ÷ 4** ▪ 1/4 ÷ 5

1/4

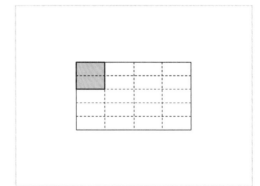

Level 3: Reasoning - Multiple division questions and misconceptions.

⚙ **Required:** 5/5   ⚙ **Student Navigation:** on
⚙ **Randomised:** off

21. What missing integer makes the following equation correct?

$1/2 \div ? = 3/18$

■ 6

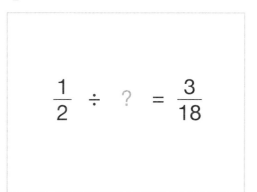

$$\frac{1}{2} \div ? = \frac{3}{18}$$

22. To cross the river, calculate the answer to each of the problems and arrange the stepping stones in ascending order (smallest first).

■ 3/10 ÷ 5 ■ 3/8 ÷ 6 ■ 2/3 ÷ 8 ■ 5/16 ÷ 3 ■ 2/3 ÷ 6

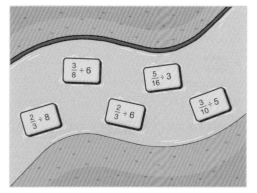

23. Which symbol makes the following statement correct?

$3/4 \div 5 \_\_ 3/7 \div 3$

1/3   ■ < ■ > ■ =

24. Robyn says, "I can't divide 5/12 by 3 because 5 is not divisible by 3". Is Robyn correct? Explain your answer.

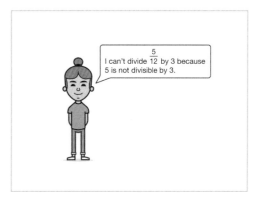

25. Which two calculations are **incorrect**?

■ 3/7 ÷ 5 = 3/35 ■ 1/4 ÷ 12 = 12/48 ■ 14/15 ÷ 8 = 7/60
■ 3/8 ÷ 12 = 1/32 ■ 15/22 ÷ 6 = 5/42

2/5

Level 4: Problem Solving - Problems involving dividing fractions by an integer.

⚙ **Required:** 5/5   ⚙ **Student Navigation:** on
⚙ **Randomised:** off

26. A rectangle has an area of 6/7 square metre (m²). The length of the rectangle is 4 metres (m). What is its height?
Give your answer in its simplest form and include the units m (metres) in your answer.

■ 3/14 metres ■ 3/14 metre ■ 3/14 m

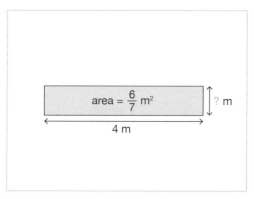

$$\text{area} = \frac{6}{7} \text{ m}^2$$
? m
4 m

27. Debbie buys a 2 litre bottle of water and drinks 350 millilitres (ml). She lets her three friends share the remaining water. If they each take an equal amount, what fraction of a litre does each friend receive?
Give your answer in its simplest form. Don't include the units in your answer.

■ 11/20

28. The heights of three plants are given in metres:
1/3 metre
1/6 metre
2/9 metre
What is their mean average height?
Include the units m (metres) in your answer.

■ 13/54 metre ■ 13/54 metres ■ 13/54 m

Level 4: *cont.*

29. Laura thinks of a fraction and performs the
following calculations on it:

+1, ×2, +3, ×4, +5, ×6

Laura's answer is 182. What was her original
fraction?

▪ **2/3**

30. In the diagram the number in the box can be
calculated by $x \div y \times z$. What is the value of y?

▪ **4**

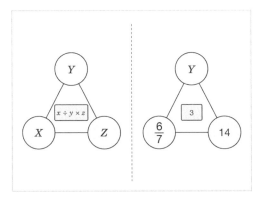

Divide Mixed Number Fractions by Integers

Competency: Use the four operations, including formal written methods, applied to integers, decimals, proper and improper fractions and mixed numbers, all both positive and negative.

Quick Search Ref: 10337

Correct: Correct.   Wrong: Incorrect, try again.   Open: Thank you.

Level 1: Understanding - Dividing a mixed number fraction by an integer, including simplifying.

✿ **Required:** 7/10   ✿ **Student Navigation:** on   ✿ **Randomised:** off

1. What is the method for dividing a mixed number fraction by an integer?

1/3

- Convert to an improper fraction then divide the denominator by the integer; the numerator stays the same.

- Convert to an improper fraction then multiply the denominator by the integer; the numerator stays the same.

- Convert to an improper fraction then multiply the numerator by the integer; the denominator stays the same.

2. When you calculate 2 2/3 ÷ 5, what will the denominator be?

■ 15

$$2\frac{2}{3} \div 5$$
$$\Downarrow \qquad \Downarrow$$
$$\frac{8}{3} \div 5 = \frac{8}{?}$$

3. When you calculate 1 3/4 ÷ 6, what will the numerator be?

■ 7

$$1\frac{3}{4} \div 6$$
$$\Downarrow \qquad \Downarrow$$
$$\frac{7}{4} \div 6 = \frac{?}{24}$$

4. What is 1 1/4 ÷ 3?

 ■ 5/12

$$1\frac{1}{4} \div 3$$
$$\Downarrow \qquad \Downarrow$$
$$\frac{?}{4} \div 3 = \frac{?}{?}$$

5. What is 1 1/3 ÷ 5?

 ■ 4/15

$$1\frac{1}{3} \div 5$$
$$\Downarrow \qquad \Downarrow$$
$$\frac{?}{3} \div 5 = \frac{?}{?}$$

6. Calculate 1 3/5 ÷ 4.
Give your answer in its simplest form.

  ■ 2/5

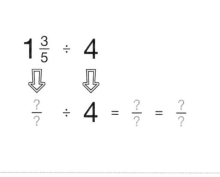

7. What is 1 1/7 ÷ 6?
Give your answer in its simplest form.

 ▪ 4/21

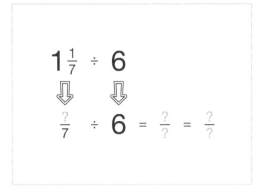

8. Calculate 1 1/6 ÷ 3.

 ▪ 7/18

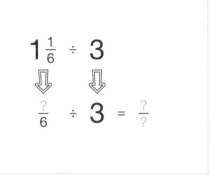

9. What is 3 5/12 ÷ 4?

 ▪ 41/48

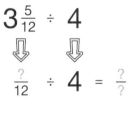

10. What is 2 1/4 ÷ 5?

 ▪ 9/20

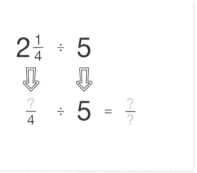

Level 2: Fluency - Dividing a mixed number fraction by an integer in context and inverse questions.

✿ **Required:** 7/10   ✿ **Student Navigation:** on
✿ **Randomised:** off

11. A recipe requires 1 1/4 of a kilogram (kg) of self-raising flour to make 8 cupcakes. How much self-raising flour do you need to make one cupcake?
Give your answer as a fraction and include the units kg (kilogram) in your answer.

▪ 5/32 kilograms ▪ 5/32 kg ▪ 5/32 kilogram

12. What missing integer makes the following statement correct?
2 2/5 ÷ ? = 12/35

▪ 7

$$2\frac{2}{5} \div \ ? \ = \frac{12}{35}$$

13. Noah makes a recipe for 9 people using 2 5/8 kilograms (kg) of chicken. How much chicken is used for each person?
Give your answer as a fraction in its simplest form. Don't include the units in your answer.

▪ 7/24

14. Calculate four and one-third divided by four.
Give your answer as a mixed number fraction in digits.

▪ 1 1/12

15. What missing mixed number fraction makes the following statement correct?

a b c

? ÷ 4 = 11/28

▪ 1 4/7

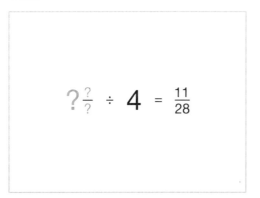

$$? \frac{?}{?} \div 4 = \frac{11}{28}$$

16. Select the two options that make the following statement true:

▢ ☒ ▢

___ ÷ ___ = 1 3/7.

2/6 　▪ 1 5/7 ▪ 3 ▪ 5 2/3 ▪ 7 ▪ 4 2/7 ▪ 6

17. 4 5/8 litres (l) of juice fills 4 identical bottles. How much juice does 1 bottle hold?

a b c

Give your answer as a mixed number fraction and include the units l (litres) in your answer.

▪ 1 5/32 litre ▪ 1 5/32 litres ▪ 1 5/32 l

18. What is three and three-fifths divided by six?

a b c

Give your answer in its simplest form.

▪ 3/5

19. Carol completes an 8 1/4 mile walk in three hours. How far does Carol walk in one hour?

a b c

Give your answer in miles as a mixed number. Don't include the units in your answer.

▪ 2 3/4

20. What missing integer makes the following statement correct?

1 2 3

4 1/7 ÷ ? = 1 8/21

▪ 3

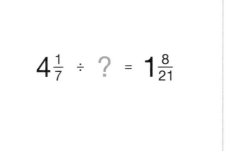

$$4\frac{1}{7} \div ? = 1\frac{8}{21}$$

21. What missing integer makes the following equation correct?

1 2 3

4 1/2 ÷ ? = 3/4

▪ 6

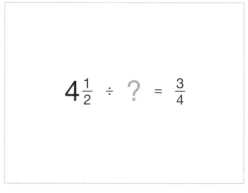

$$4\frac{1}{2} \div ? = \frac{3}{4}$$

22. Which symbol makes the following statement correct?

▢ ☒ ▢

4 5/9 ÷ 8 ___ 5 1/4 ÷ 9

1/3 　▪ < ▪ > ▪ =

23. Georgia says, "I can't divide 2 7/12 by 5 because 12 is not divisible by 5". Is Georgia correct? Explain your answer.

a b c

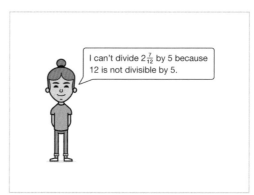

I can't divide $2\frac{7}{12}$ by 5 because 12 is not divisible by 5.

24. Which two calculations are **incorrect**?

▢ ☒ ▢

▪ 3/7 ÷ 5 = 3/35 ▪ 14/15 ÷ 8 = 7/60 ▪ 15/22 ÷ 6 = 5/42
▪ 1 7/8 ÷ 12 = 5/32 ▪ 5 1/4 ÷ 12 = 5/12

2/5

25. Jonny calculates 6 2/9 ÷ 3 and gets the answer 6 2/27. Explain the mistake he has made.

a b c

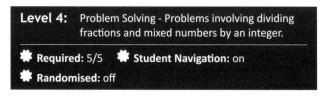

Level 4:   Problem Solving - Problems involving dividing
fractions and mixed numbers by an integer.

⚙ **Required:** 5/5   ⚙ **Student Navigation:** on
⚙ **Randomised:** off

26. Arrange the calculations in ascending order
(smallest first).

■ `1 7/8 ÷ 3` ■ `3 1/5 ÷ 5` ■ `5 1/3 ÷ 8` ■ `4 1/2 ÷ 6`
■ `2 4/7 ÷ 3`

27. A rectangular garden has an area of 115.5 square
metres (m²). The length of the garden is 14 metres
(m). What is the width?
*Give your answer as a mixed number fraction and
include the units m (metres) in your answer.*

■ `8 1/4 metres` ■ `8 1/4 m` ■ `8 1/4 metre`

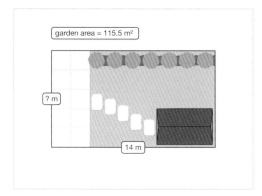

garden area = 115.5 m²

? m

14 m

28. Debbie buys a 2 litre bottle of water and drinks
350 millilitres (ml). She lets her three friends share
the remaining water. If they each take an equal
amount, what fraction of a litre does each friend
receive?
*Give your answer in its simplest form. Don't
include the units in your answer.*

■ `11/20`

29. The heights of three trees are given in metres:
1 1/2 metres
2 1/4 metres
1 2/3 metres
What is their mean average height?
*Give your answer as a mixed number fraction in its
simplest form and include the units m (metres) in
your answer.*

■ `1 29/36 m` ■ `1 29/36 metre` ■ `1 29/36 metres`

30. Riley thinks of a mixed number fraction
and performs the following calculations on it:
×3, ×5, -1, +5, ×7
Riley's answer is 154. What was his original mixed
number fraction?

■ `1 1/5`

$\frac{2}{7}$ ×3 ×5 -1 +5 ×7 = 154

Find a fractional increase or decrease

Competency: Use the four operations, including formal written methods, applied to integers, decimals, proper and improper fractions and mixed numbers, all both positive and negative.

Quick Search Ref: 10147

Correct: Correct.   **Wrong:** Incorrect, try again.   **Open:** Thank you.

Level 1: Understanding - Increasing and decreasing by unit fractions.

✱ **Required:** 7/10   ✱ **Student Navigation:** on   ✱ **Randomised:** off

---

1. Sort the following steps in order to find a fractional increase.

↑↓

■ Divide the amount by the denominator.

■ Multiply your answer by the numerator.

■ Add your answer to the original amount.

2. Which calculation would you use to increase 70 by 1/5?

▣
1/4

■ 70 - (70 ÷ 5) ■ 70 + (70 ÷ 5) ■ 70 ÷ 5 ■ 70 + (70 × 5)

3. Increase 35 by 1/5.

123 ■ 42

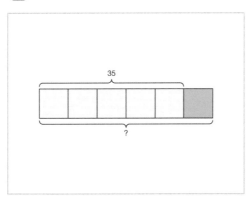

4. Increase 42 by 1/3.

123 ■ 56

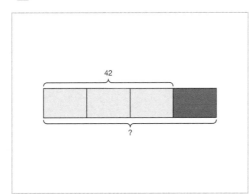

5. Decrease 36 by 1/4.

123 ■ 27

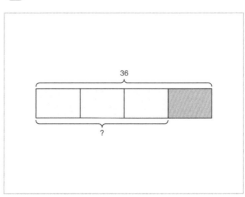

6. Decrease 84 kilograms (kg) by 1/7.
Include the units kg (kilograms) in your answer.

abc ■ 72 kg ■ 72 kilograms

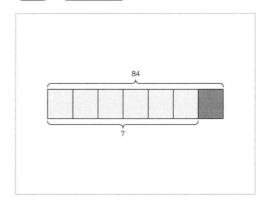

7. Increase 72 metres (m) by 1/8.
Include the units m (metres) in your answer.

abc ■ 81 metres ■ 81 m

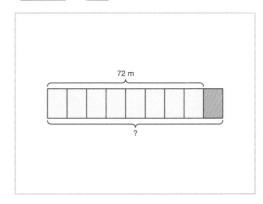

Level 1: *cont.*

8. Increase 48 by 1/6.

 ▪ 56

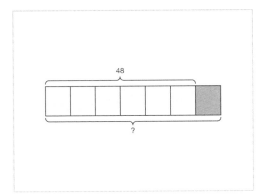

9. Decrease 54 litres (l) by 1/9.
Include the units l (litres) in your answer.

 ▪ 48 l ▪ 48 litres

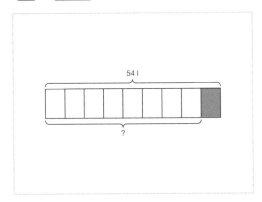

10. Increase 108 kilometres (km) by 1/12.
Include the units km (kilometres) in your answer.

 ▪ 117 kilometres ▪ 117 km

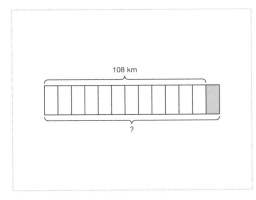

Level 2: Fluency - Increasing and decreasing by proper fractions in context.

✿ **Required:** 7/10   ✿ **Student Navigation:** on
✿ **Randomised:** off

11. Increase 18 by 2/3.

 ▪ 30

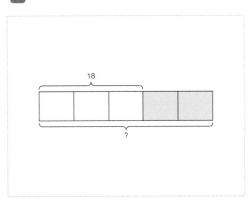

12. Increase 64 by 5/8.

 ▪ 104

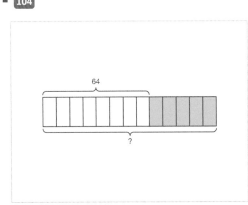

13. Decrease 280 by 3/5.

 ▪ 112

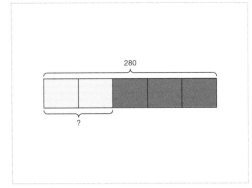

14. Decrease 1.54 kilometres (km) by 4/7.
Include the units (m) in your answer.

a b c

- 660 metres - 660 m

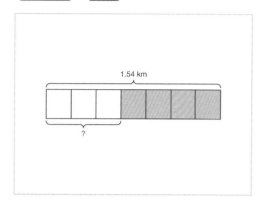

15. Last season, a football team scored 72 goals. This season the number of goals they scored increased by 2/9. How many goals did the team score this season?

1 2 3

- 88

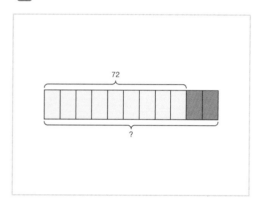

16. Olivia subscribed to a cheaper mobile phone plan to reduce her monthly bill by three-eights. If the last bill on her old plan was £30.24, what will her new bill be?
Include the £ sign in your answer.

a b c

- £18.90

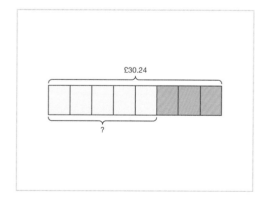

17. A car factory produced 120,000 cars last year. This year they plan to increase their production by two-fifths. How many cars will they produce this year?

a b c

- 168,000 - 168000

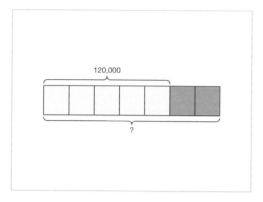

18. A farmer predicts that her wheat crop will decrease by 3/7 this year due to bad weather. If she harvested 56 tonnes of wheat last year, how many tonnes does she expect to harvest this year?
Don't include units in your answer.

1 2 3

- 32

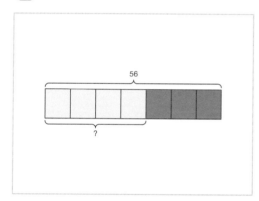

19. The cost of a train ticket has been reduced by one-quarter. If the original ticket cost £51, what is the new price?
Include the £ sign in your answer.

a b c

- £38.25

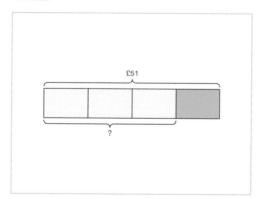

20. Increase 174 kilograms (kg) by 5/6.
a b c *Include the units kg (kilograms) in your answer.*

■ 319 kg ■ 319 kilograms

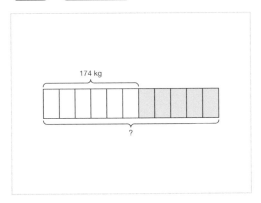

25. A designer sweatshirt originally cost £48 but Calvin
a b c bought it in a sale for £28.80. What fractional
discount did he receive?

■ 2/5

Level 3: Reasoning - Using multipliers to find fractional
increases and decreases.

❋ **Required:** 5/5 ❋ **Student Navigation:** on
❋ **Randomised:** off

Level 4: Problem Solving - Multi-step problems with
fractional increase and decreases.

❋ **Required:** 5/5 ❋ **Student Navigation:** on
❋ **Randomised:** off

21. Amy says that decreasing an amount by one-third
a b c is the same as multiplying the amount by two-
thirds.
Is Amy correct? Explain your answer.

22. What is the difference between the sale price of
a b c jeans at Best Fit and at Right Value?
Include the units p (pence) in your answer.

■ 50p

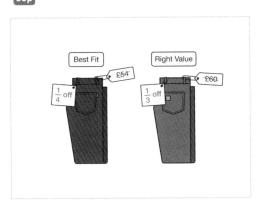

23. Carlos says, 'Increasing an amount by 2/5 is the
a b c same as multiplying the amount by 7/5'.
Is Carlos correct? Explain your answer.

24. The height of a sunflower increases by two-fifths
a b c in one month. If the sunflower is now 35
centimetres (cm) tall, how tall was it a month ago?
Include the units cm (centimetres) in your answer.

■ 25 centimetres ■ 25 cm

26. 48 increased by 1/6 is the same as ___ reduced by
a b c 1/9.

■ 63

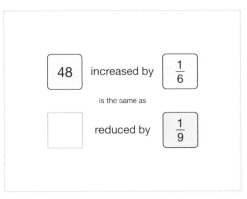

27. The population of Sumtown increases by one-third
a b c each year. If the population is now 128,000 what
was the population three years ago?

■ 54,000 ■ 54000

28. If you increase the length of each side of a square
a b c by one-quarter, what fraction does the area
increase by?

■ 9/16

29. There is a full box of apples on the left of the
a scales and a partially filled box on the right. When
b 3/10 of the apples are moved from the left to the
c right, the scales balance. What has been the
fractional increase in the amount of apples on the
right?

- 3/4

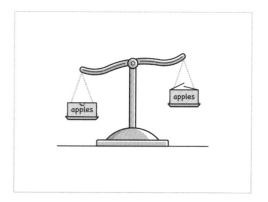

30. The price of a car decreases by 2/9 each year. If
a the current value of the car is £9,800, what was its
b value two years ago?
c *Include the £ sign in your answer.*

- £16,200 - £16200

Find the whole amount given a fraction of the amount

Competency: Use the four operations, including formal written methods, applied to integers, decimals, proper and improper fractions and mixed numbers, all both positive and negative.

Quick Search Ref: 10254

Correct: Correct.   Wrong: Incorrect, try again.   Open: Thank you.

Level 1: Understanding - Find a fraction of an amount (unit fractions and proper fractions)

⚙ Required: 7/10   ⚙ Student Navigation: on   ⚙ Randomised: off

---

1. What is 1/3 of 18?

▪ 6

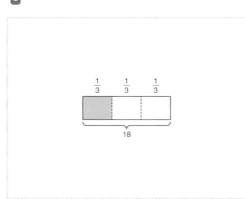

2. Find 1/7 of 56.

▪ 8

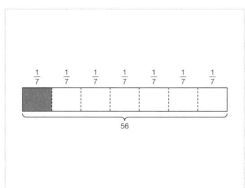

3. Calculate 3/5 of 40.

▪ 24

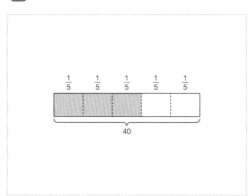

4. What is 5/12 of 84?

▪ 35

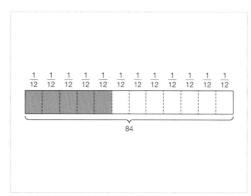

5. Find 5/8 of £6.00.
Give your answer in pounds and pence.
Include the £ sign in your answer.

▪ £3.75

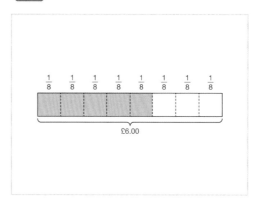

6. Calculate 2/3 of 20.
Give your answer as a mixed number fraction.

▪ 13 1/3

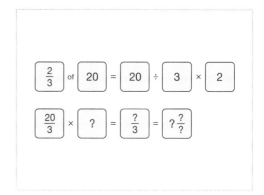

Level 1: *cont.*

7. Kayleigh has 7 metres (m) of ribbon and uses 3/4
a of it to wrap presents. How much ribbon does
b Kayleigh use?
c Give your answer as a decimal number.
Include the units m (metres) in your answer.

▪ 5.25 metres ▪ 5.25 m

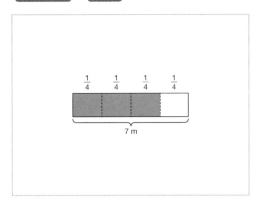

8. Calculate 5/9 of 108.
1
2 ▪ 60
3

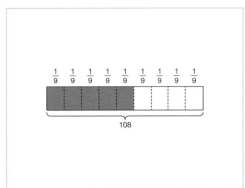

9. A school has 720 students and two-fifths have
1 school dinners. How many students have school
2 dinners?
3

▪ 288

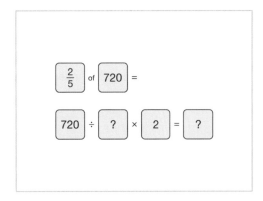

10. What is 3/7 of 30?
a Give your answer as a mixed number fraction.
b
c ▪ 12 6/7

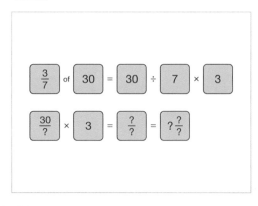

Level 2: Fluency - Find the whole amount given a
fraction of the amount.
─────────────────────────────
✴ Required: 7/10   ✴ Student Navigation: on
✴ Randomised: off

11. 1/3 of an amount is 5, what is the whole amount?
1
2 ▪ 15
3

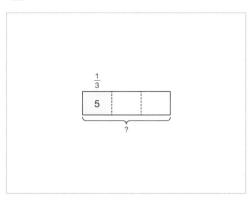

12. 1/6 of an amount is 8, what is the whole amount?
1
2 ▪ 48
3

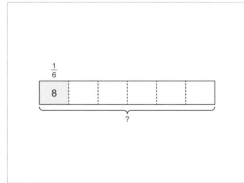

13. 1/7 of a year group chose Maths as their favourite subject. If 28 people choose Maths, how many are there in the year group?

1 2 3

▪ 196

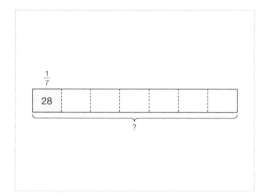

14. If 2/5 of an amount is 20, what is the amount?

a b c

▪ 50

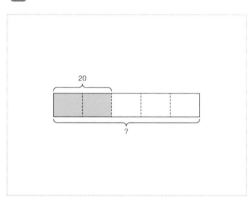

15. Soloman gives 3/8 of his trading cards to Trevor. If Trevor receives 72 cards, how many cards did Soloman start with?

a b c

▪ 192

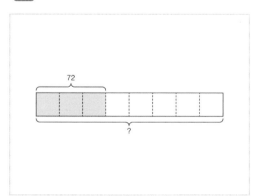

16. What missing value makes the statement true?
9/14 of ___ = 36.

a b c

▪ 56

17. Sofia washes 7 cars and Noman washes 5 cars. They decide to split their earnings according to how many cars they each washed. If Noman earns £32, how much money did they earn altogether? *Include the £ sign in your answer.*

a b c

▪ £76.80

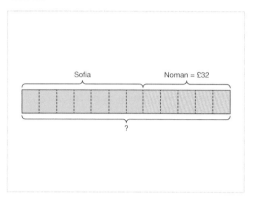

18. A driver has completed 1/6 of a race. If the driver has completed 9 laps, how many laps are in the race altogether?

1 2 3

▪ 54

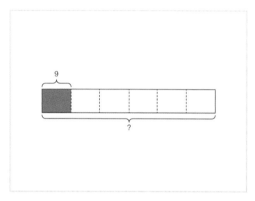

19. The price of a coat is reduced in a sale by 2/5. If the sale price of the coat is £24, what was the original price of the coat? *Include the £ sign in your answer.*

a b c

▪ £40 ▪ £40.00

20. What missing value makes the statement true?
2/7 of ___ = 42.

1 2 3

▪ 147

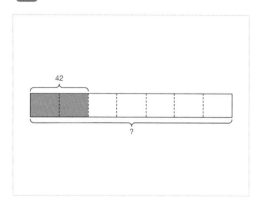

Level 3:   Reasoning - Two-step calculations.

✱ **Required:** 5/5   ✱ **Student Navigation:** on
✱ **Randomised:** off

21. Select the symbol that makes the following statement true.

☐
☒   7/12 of £3,000,000 ____ 3/8 of £5,000,000.
☐

1/3   ▪ < ▪ = ▪ >

22. David is trying to find the whole amount when 2/3 of the amount is £48.
Here is David's working. Is he correct?
Explain your answer.

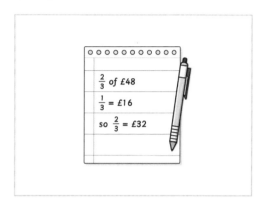

23. One-twelfth of an amount is 9. What is one-sixth of the amount?

▪ 18

24. 3/4 of A = 24, 2/3 of B = 20.
Find the value of A - B.

▪ 2

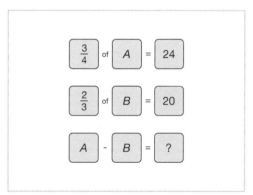

25. A school sells 5/8 of its tickets for a raffle. If there are 360 tickets left to sell, how many tickets were there originally?

▪ 960

Level 4:   Problem Solving - Multi-step problems.

✱ **Required:** 5/5   ✱ **Student Navigation:** on
✱ **Randomised:** off

26. 2/3 of a class are boys and 3/5 of the boys choose football as a winter sport. If twelve boys choose football, how many students are in the class?

▪ 30

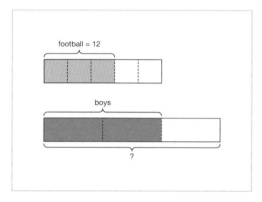

27. A bus is one-fifth full. Three people get on and the bus is one-quarter full.
How many people can the bus hold?

▪ 60

28. Robert gives 1/3 of his cards to Sara and then 2/5 of what's left to Tulisa.
If Tulisa gets twelve cards, how many cards did Robert begin with?

▪ 45

29. Every student in Year 8 must choose to study French, Spanish or both. Two-thirds choose Spanish and 70% choose French. If 55 students choose both languages, how many students are in Year 8?

▪ 150

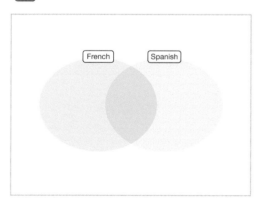

Level 4: cont.

30. One-fifth of square (a) overlaps one-quarter of square (b).

1
2
3

If the area of the combined shape is 480 cm² (square centimetres), what is the area of square (b)?

Don't include units in your answer.

▪ **240**

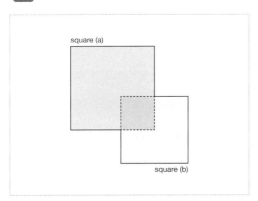

Mathematics

Decimals and Percentages

Percentages

Equivalence

Calculate the percentage change between two amounts

Competency: Solve problems involving percentage change, including: percentage increase, decrease and simple interest in financial mathematics.

Quick Search Ref: 10205

Correct: Correct.   Wrong: Incorrect, try again.   Open: Thank you.

Level 1: Understanding - Calculate the percentage change between two amounts.

✿ **Required:** 7/10   ✿ **Student Navigation:** on   ✿ **Randomised:** off

---

1. If a number is 150% of the original number, what is the percentage change?

1/4

- **50% increase** ▪ 50% decrease ▪ **150% increase**
- **150% decrease**

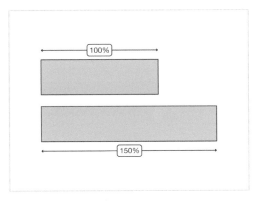

2. If a number is 70% of the original number, what is the percentage change?

1/4

- **70% increase** ▪ **70% decrease** ▪ **30% increase**
- **30% decrease**

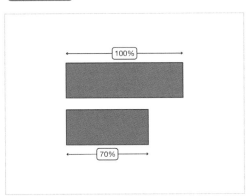

3. What is **40** as a percentage of **50**?

- **80%** ▪ **80**

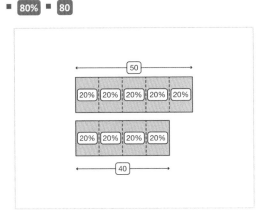

4. What is 40 as a percentage of 32?

- **125%** ▪ **125**

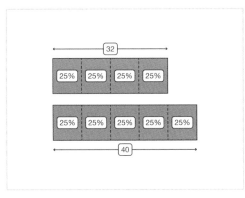

5. Which calculation would you use to find the percentage change when 10 is increased to 15?

1/4

- ▪ 10 ÷ 15 x 100 -100 ▪ **15 ÷ 10 x 100 - 100**
- ▪ 100 - 15 ÷ 10 x 100 ▪ 100 - 10 ÷ 15 x 100

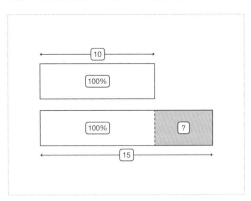

6. What is the percentage change when 16 is reduced to 7.2?
Select **either** decrease or increase **and** the correct percentage.

2/5

- ▪ **Decrease** ▪ Increase ▪ 22% ▪ **55%** ▪ 45%

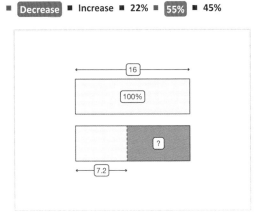

Level 1: cont.

7. A number is increased from 18 to 29.16. What is the percentage change?
Select **either** decrease or increase **and** the correct percentage.

2/5

- Decrease ▪ **Increase** ▪ 162% ▪ 38% ▪ **62%**

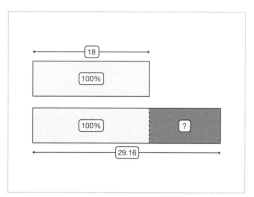

8. Which calculation would you use to find the percentage change when 20 is decreased to 15?

1/4

- **15 ÷ 20 x 100 -100** ▪ 20 ÷ 15 x 100 - 100
- 100 - 15 ÷ 20 x 100 ▪ 100 - 20 ÷ 15 x 100

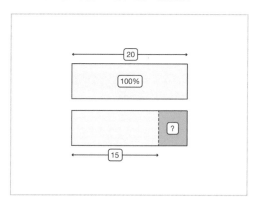

9. A number is reduced from 36 to 26.28. Calculate the percentage change.
Select **either** decrease or increase **and** the correct percentage.

2/5

- **Decrease** ▪ Increase ▪ 37% ▪ 73% ▪ **27%**

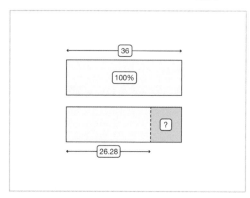

10. What is the percentage change when 56 is increased to 104.16?
Select **either** decrease or increase **and** the correct percentage.

2/5

- **Increase** ▪ Decrease ▪ **86** ▪ 186 ▪ 46

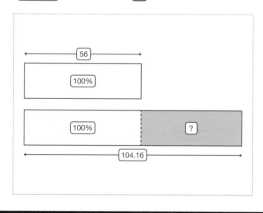

Level 2: Fluency - Calculate the percentage change between two amounts given in worded problems.

✿ **Required:** 7/10   ✿ **Student Navigation:** on
✿ **Randomised:** off

11. In 2010 there were 3,200 tigers in the wild and in 2016 there were 3,890.
What is the percentage increase?
Give your answer to the nearest whole number.

a
b
c

- **22%** ▪ **22**

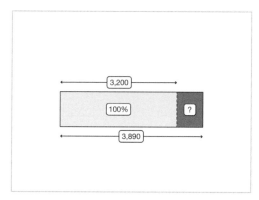

12. A car was bought for £15,880 and it was sold on for £10,800.
What was the percentage decrease in value?
Give your answer to the nearest whole number.

a
b
c

- **32%** ▪ **32**

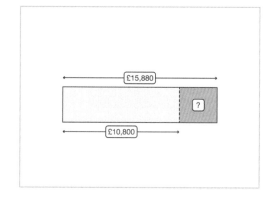

13. A house was bought in 1997 for £79,950.
In 2017 it was sold for £175,000.
By how many percent did the price increase?
Give your answer to the nearest whole number.

a
b
c

■ 119% ■ 119

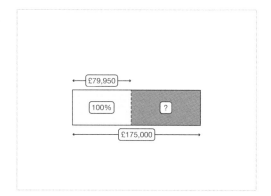

14. In Supermarket A, 15 grocery items cost £14.75.
The same 15 items cost £16.62 in Supermarket B.
As a percentage, what do you save shopping by in Supermarket A?
Give your answer to the nearest whole number.

a
b
c

■ 11% ■ 11

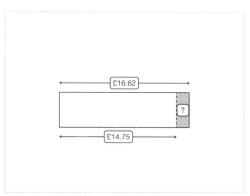

15. It costs a restaurant £3.20 to make a portion of lasagne, which it sells for £9.95. What percentage profit does the restaurant make? Give your answer to the nearest number.

a
b
c

■ 211% ■ 211

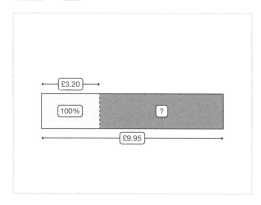

16. In 2007 Usain Bolt ran 100 metres in 10.03 seconds.
In 2008 he ran the same distance in 9.69 seconds.
What was the percentage decrease in his time?
Give your answer to the nearest whole percent.

a
b
c

■ 3% ■ 3

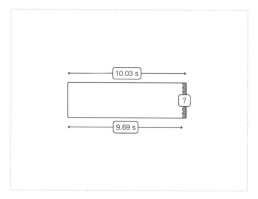

17. A console game costing £19.99 is reduced in the sale to £17.34.
By how many percent has it been reduced?
Give your answer to the nearest whole percent.

a
b
c

■ 13% ■ 13

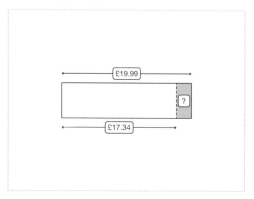

18. In July the temperature rises from 18.6 °C to 22.6 °C.
What is the percentage increase in temperature?
Give your answer to the nearest whole percent.

a
b
c

■ 22 ■ 22%

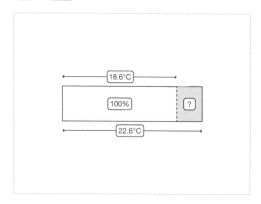

Level 2: *cont.*

19. Harvey is making a cake. He increases the amount
a of sugar in the recipe from 450 grams to 525
b grams.
c By what percentage has he increased the sugar?
Give your answer to the nearest whole percent.

▪ 17% ▪ 17

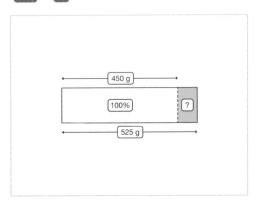

20. Alice's salary is increased from £22,200 to
1 £23,976. What is her percentage pay rise?
2 Give your answer to the nearest whole percent.
3

▪ 8

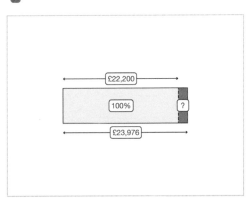

Level 3: Reasoning - Interpreting percentage change
between two amounts within a question.

✻ **Required:** 5/5   ✻ **Student Navigation:** on
✻ **Randomised:** off

21. If *x* is increased to 5*x*, what is the percentage
a increase?
b
c ▪ 400 ▪ 400%

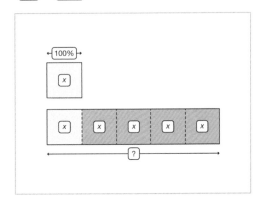

22. If *y* is decreased to 0.8*y* what is the percentage
1 decrease?
2
3 ▪ 20

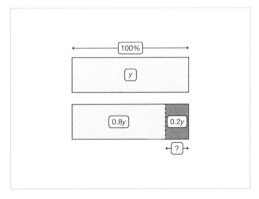

23. A quantity is increased from 5 to 7.
a To find the percentage increase, Sally calculates:
b 7/5 x 100 -100.
c Ali calculates: 2/5 x 100.
Will both methods give the correct answer?
Explain your answer.

24. Arrange the following pay rises in order (largest
↑ percentage pay rise first).
↓
▪ £64,000 to £65,024 ▪ £13,000 to £13,195
▪ £34,000 to £34,476 ▪ £27,500 to £27,830

25. A number has been increased to 20.4.
1 If the percentage increase was 70%, what was the
2 original number?
3
▪ 12

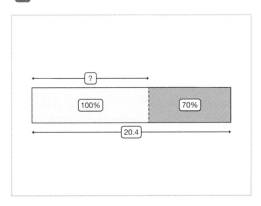

Level 4: Problem Solving - Using percentage change
between two amounts to solve complex
problems.

✻ **Required:** 5/5   ✻ **Student Navigation:** on
✻ **Randomised:** off

26. Mrs Kinder orders 9 boxes of purple pens.
a Each box of 20 pens costs £4.02 plus a £5 delivery
b charge.
c Mrs Kinder sells all of the pens for 20 pence each.
Work out the percentage loss on her total cost to
the nearest whole percent.

▪ 13 ▪ 13%

Level 4: *cont.*

27. A 40 gram bar of chocolate costs 60 pence. If the
weight of the bar is increased to 43.2 grams and
the cost of the bar is increased by the same
percentage, what is the new cost of the chocolate
bar in pence?
Include the p (pence) symbol in your answer.

a
b
c

▪ 65p ▪ 65 ▪ 65 pence

28. Evie runs the London Marathon in 4:45:30.
The following month she completes the Edinburgh
Marathon in 4:06:20.
Calculate the percentage decrease in her time.
Give your answer to the nearest whole number.

a
b
c

▪ 14 ▪ 14%

29. Jessica uses 4/7 of a roll of fabric to cover a chair.
If she reduces the amount of fabric she uses to 3/8
of a roll for the next chair she covers, what is the
percentage reduction in the amount of fabric
used?

a
b
c

▪ 34 3/8% ▪ 34.375% ▪ 34 3/8 ▪ 34.375

30. Abbas and Fatima compare the household chores
they do each week.
The week one ratio for Abbas:Fatima is 2:3.
The week two ratio is 5:4.
By what percentage does Abbas increase his
number of chores from week one to week two?
Give your answer to the nearest whole number.

a
b
c

▪ 39 ▪ 39%

Compare two quantities using percentages

Competency: define percentage as 'number of parts per hundred', interpret percentages and percentage changes as a fraction or a decimal, interpret these multiplicatively, express one quantity as a percentage of another, compare two quantities using percentages, and work with percentages greater than 100%

Quick Search Ref: 10236

Correct: Correct.   **Wrong:** Incorrect, try again.   **Open:** Thank you.

Level 1: Understanding - Convert quantities into percentages and compare them without using a calculator.

✹ **Required:** 7/10   ✹ **Student Navigation:** on   ✹ **Randomised:** off

1. Give 30 out of 40 as a fraction.

a b c   ▪ 3/4 ▪ 30/40

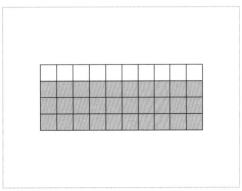

2. What is 3/5 as a percentage?

1 2 3   ▪ 60

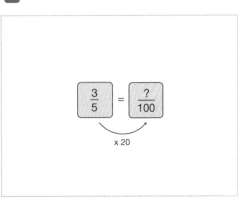

3. Give 28 out of 35 as a percentage.

a b c   ▪ 80

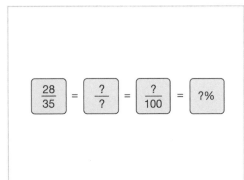

4. What is 35 as a percentage of 28?

1 2 3   ▪ 125

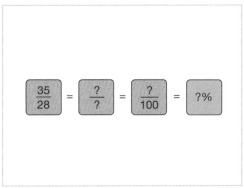

5. What is the percentage difference between **8 out of 32** and **34%**?

1 2 3   ▪ 9

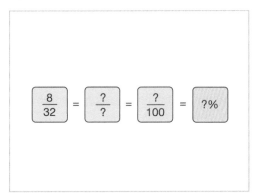

6. Which is greater:

☐ ☒ ☐ ☐   ▪ 13 out of 20 ▪ 16 out of 25

1/2

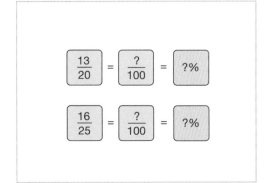

Level 1: *cont.*

7. "14 out of 70 is greater than 7 out of 28".
Is this statement true or false?

■ True ■ False

1/2

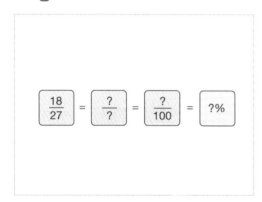

8. What symbol makes the statement true?

33% ____ 18 out of 27

1/4 ■ > ■ < ■ = ■ ≥

$$\frac{18}{27} = \frac{?}{?} = \frac{?}{100} = ?\%$$

9. Compare the two values and give the smallest
value as a percentage.
9 out of 45
7 out of 28

■ 20

$$\frac{9}{45} = \frac{?}{?} = \frac{?}{100} = ?\%$$

$$\frac{7}{28} = \frac{?}{?} = \frac{?}{100} = ?\%$$

10. Compare the two values and give the largest value
as a percentage.
24 out of 30
15 out of 20

■ 80

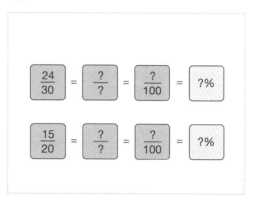

Level 2: Fluency - Finding percentages and comparing
them using a calculator including percentages
greater than 100.

🌸 **Required:** 7/10   🌸 **Student Navigation:** on
🌸 **Randomised:** off

11. Give 14 out of 48 as a percentage to the nearest
whole number.

■ 29

$$14 \div 48 =$$

12. What is 58 as a percentage of 32?
Give your answer to the nearest whole number.

■ 181

13. A 0.85 kg tin of chocolates contains 192 grams of
fat.
What percentage of the chocolates is fat?
Give your answer to the nearest whole number.

■ 23

14. Select the symbol that makes the following
statement true:
180 out of 236 _____ 480 out of 645

1/4 ■ < ■ > ■ = ■ ≤

15. Rosie scores 24 out of 60 in her first test and 12
out of 40 in her second test. What is the difference
in her scores as a percentage?

■ 10

Level 2: *cont.*

16. Cerys has 8 pieces of a 240 piece jigsaw left to complete.
Beth has 15 pieces of a 480 piece jigsaw left to complete.

1/2 Who has completed the larger proportion of their jigsaw?

▪ **Beth** ▪ **Cerys**

17. An orange weighing 184 grams contains 16.56 grams of sugar.
A slice of melon weighing 125 grams contains 10 grams of sugar.
What is the percentage difference in sugar content?

▪ **1**

18. Driving school A has 386 learner drivers and 284 pass their test first time.
Driving school B has 220 learner drivers and 154 pass their test first time.
What is the percentage pass rate of the more successful school? Give your answer to the nearest percentage.

▪ **74**

19. Which of the following symbols makes the statement true?
122 out of 380 _____ 148 out of 463

1/5 ▪ **>** ▪ ≤ ▪ = ▪ < ▪ ≥

20. A 42 gram bar of chocolate contains 12.3 grams of fat.
A 37 gram bag of chocolate contains 9.1 grams of fat.
What is the larger percentage fat content? Give your answer to the nearest percent.

▪ **29**

Level 3: Reasoning - comparing percentage of amounts to make decisions.

✿ **Required:** 5/5 ✿ **Student Navigation:** on
✿ **Randomised:** off

21. Aisha makes 3 glasses of orange. Arrange them in order of strength starting with the glass that contains the greatest percentage of cordial:
Glass a: 100 ml orange cordial, 300 ml water.
Glass b: 136 ml orange cordial, 410 ml water.
Glass c: 128 ml of orange cordial, 380 ml water.

▪ **C** ▪ **A** ▪ **B**

22. 9 carat gold contains exactly 38% solid gold. Which of the following items is 9 carat gold?

1/3 ▪ **Necklace** ▪ **Ring** ▪ **Bracelet**

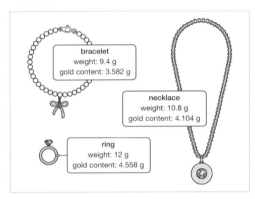

bracelet
weight: 9.4 g
gold content: 3.582 g

necklace
weight: 10.8 g
gold content: 4.104 g

ring
weight: 12 g
gold content: 4.558 g

23. Matthew gets 34 out of 40 on the first part of his exam and 26 out of 48 on the second part of his exam.
He says that to work out his total percentage for the whole paper, he can add the two percentages together and then divide by 2. Explain why he is wrong.

24. "If you want to improve your test score by 10% you need to get 10 more marks on your next test?"
Is the statement true?

1/3 ▪ **Always true** ▪ **Sometimes true** ▪ **Never true**

25. Olivia is on holiday in England and spends £14 out of the £20 she takes with her.
James is on holiday in America and spends $28 out of the $40 he takes with him.

1/2

True or false? To compare the percentage of money spent you need to start by changing the amounts into the same currency.

▪ **True** ▪ **False**

26. A cube is made up of 125 blue centimetre cubes. The outside of the cube is painted red. What percentage of the total number of cubes have been painted red?
Give your answer correct to one decimal place.

▪ 78.4

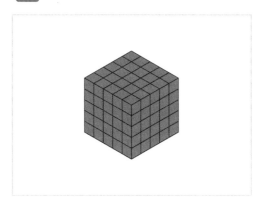

27. The diagram shows two containers, a cuboid and a cube. The cube was filled with water which was then poured into the cuboid. What percentage of the cuboid is now filled with water?
Give your answer to the nearest whole percent.

▪ 79

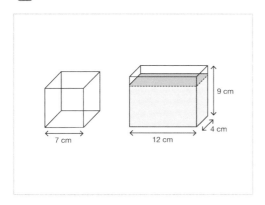

28. The ratio of girls to boys in a class is 3:2. 40% of the girls have brown eyes and 70% of the boys have brown eyes. What percentage of the pupils in the class have brown eyes?

▪ 52

29. The local swimming pool has been re-tiled. Dark blue tiles cover the floor and light blue tiles are fixed on all four sides. What percentage of the **total amount of tiles** are **dark blue**?
Give your answer as a percentage rounded to one decimal place.

▪ 70.4

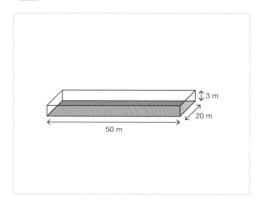

30. The diagram shows three squares made up from unit squares. In each square, the bottom row of unit squares is shaded and then one less unit square is shaded in each row above.
What percentage of a 100 by 100 square will be shaded?

▪ 50.5

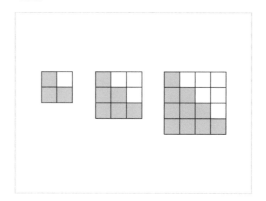

Decrease an amount by a percentage.

Competency: Solve problems involving percentage change, including: percentage increase, decrease and simple interest in financial mathematics.

Quick Search Ref: 10173

Correct: Correct.   Wrong: Incorrect, try again.   Open: Thank you.

Level 1: Understanding - How to decrease an amount by a percentage.

✿ **Required:** 7/10   ✿ **Student Navigation:** on   ✿ **Randomised:** off

1. If you decrease an amount by 20%, what percentage do you have left?

    a
    b
    c   ▪ 80 ▪ 80%

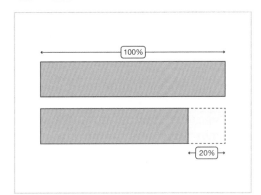

2. If you decrease an amount by 7%, what percentage do you have left?

    a
    b
    c   ▪ 93 ▪ 93%

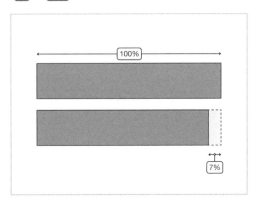

3. Express 4% as a decimal.

    1
    2
    3   ▪ 0.04

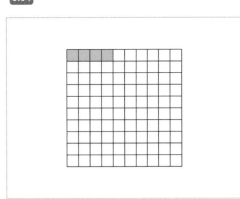

4. Select the calculation you would use to find 8% of an amount.

    ☐
    ☒
    ☐   ▪ x 0.08 ▪ x 8 ▪ x 0.8

    1/3

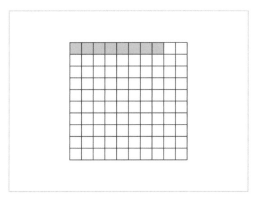

5. What would you multiply an amount by to **decrease** by 97%?

    1
    2
    3   ▪ 0.03

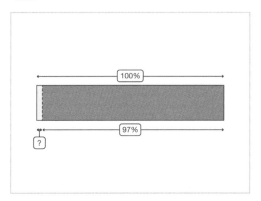

6. Select the two calculations you could use to **decrease** 54 by 62%.

    ☐
    ☒
    ☐   ▪ 54 x 0.62 ▪ 54 x 0.38 ▪ 54 - (54 x 0.62)

    2/3

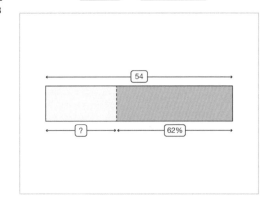

Level 1: *cont.*

7. Select the two calculations you could you use to **decrease** 44 by 4%?

■ 44 - (44 x 0.04) ■ 44 x 0.4 ■ 44 x 0.04 ■ 44 x 0.96

2/4

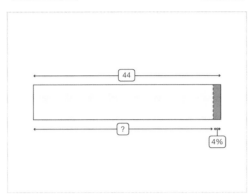

8. Select the calculation you would use to find 70% of an amount.

■ x 0.7 ■ x 7 ■ x 0.07

1/3

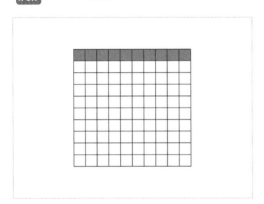

9. What would you multiply an amount by to **decrease** by 28%?

■ 0.72

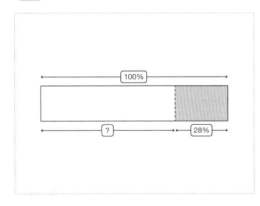

10. Select the two calculations you could use to **decrease** 83 by 12%.

■ 83 x 0.12 ■ 83 x 0.88 ■ 83 - (83 x 0.12)

2/3

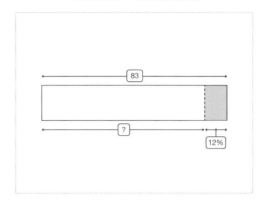

Level 2: Fluency - Calculate a percentage decrease including worded problems.

❈ Required: 7/10   ❈ Student Navigation: on
❈ Randomised: off

11. Find the new amount when 36 is decreased by 68%.

■ 11.52

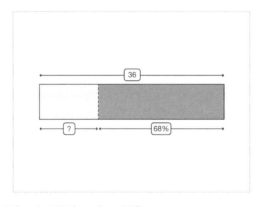

12. What is 85% less than 32?

■ 4.8

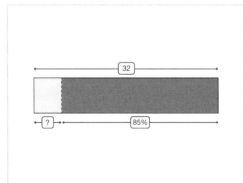

Level 2: *cont.*

13. What is 21% less than 84?

　■ 66.36

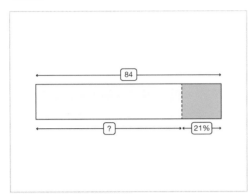

14. A restaurant offers a 13% discount on all of its food. What is the new total after the discount has been applied to a bill of £39?
Give your answer in pounds and pence.
Include the £ sign in your answer.

■ £33.93

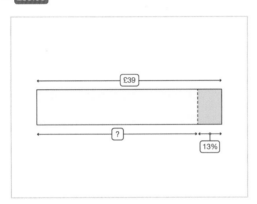

15. Reuben earns £21,450 a year.
He pays 12% National Insurance.
How much does he earn after the National Insurance is deducted?
Give your answer in pounds.
Include the £ sign in your answer.

■ £18876　■ £18,876

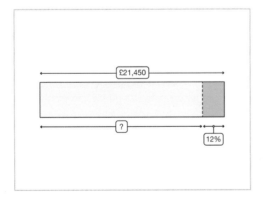

16. Last month Layla ate 16 chocolate bars and this month she wants to cut down by 17%.
How many chocolate bars can she eat this month?
Give your answer to the nearest whole number.

■ 13

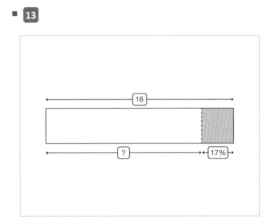

17. Freddie's end of year maths result has decreased by 13% since last year.
If he scored 86% last year, how many percent did he get this year?
Give your answer to the nearest whole number.

■ 75%　■ 75

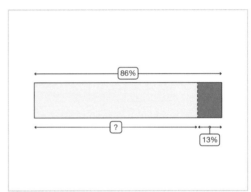

18. What is 37% less than 154?

■ 97.02

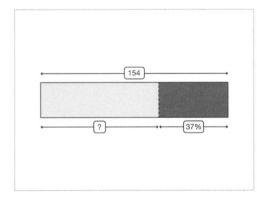

19. Find the new amount when 63 is decreased by 18%.

▪ 51.66

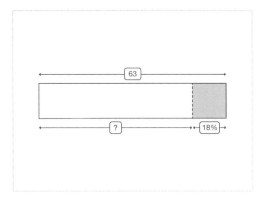

20. In Manchester, the average temperature in January was 70% lower than the average temperature in August.
If the average temperature in August is 20 °C, what is the average temperature in °C in January?

▪ 6

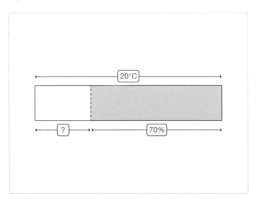

Level 3: Reasoning - Interpreting percentage decrease within a question.

✱ **Required:** 5/5   ✱ **Student Navigation:** on
✱ **Randomised:** off

21. If an amount is divided by 4, by what percentage has it been decreased?

▪ 75% ▪ 75

22. An amount has been decreased by 80%.
What do you need to multiply the **new value** by to get the original amount.

▪ 5

23. What do you need to multiply by if you want to decrease an amount **to 20%** of its original value?

▪ 0.2

24. Everything in a shop has been reduced by 15%. If the new price of an item is £20.40, what was its original price?
Give your answer in pounds.
Include the £ sign in your answer.

▪ £24 ▪ £24.00

25. A number decreases by 25% each year.
By how many percent will it have decreased after 3 years?
Give your answer to the nearest whole number.

▪ 58 ▪ 58%

Level 4: Problem Solving - Using percentage decrease to solve complex problems.

✱ **Required:** 5/5   ✱ **Student Navigation:** on
✱ **Randomised:** off

26. A farmer has a vegetable patch 35 metres long and 8 metres wide.
If he then increases the length to 40 metres, by what percentage must he decrease the width to keep the area the same?

▪ 12.5% ▪ 12.5

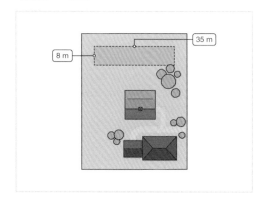

27. It's estimated that 8% of Africa's elephants are lost every year to poaching. What percentage of Africa's elephants will be lost to poaching over 3 years?
Give your answer to the nearest whole number.

▪ 22

Level 4: *cont.*

28. The cuboid has a volume of 4,800 cm³.

a Side x is decreased by 32%.

b Side y is decreased by 2%.

c Side z is decreased by 19%.

What is the volume of the new cuboid?

Give your answer to the nearest cubic centimetre.

Don't include the units in your answer.

▪ 2,591 ▪ 2591

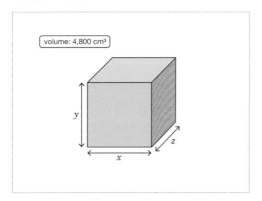

29. Savannah buys a new car. In the first year the

a value of the car depreciates by 20% and in the

b second year it depreciates by a further 14%.

c If the car is worth £8,256. at the end of the second year, what did Savannah pay for the car?

Include the £ sign in your answer.

▪ £12,000 ▪ £12000

30. In January, a shop reduces its prices by 25%.

a In February, the shop reduces everything to 50%

b of its **original price**.

c What is the percentage decrease in prices from January to February?

Give your answer to the nearest whole number.

▪ 33 ▪ 33%

Express one quantity as a percentage of another

Competency: Define percentage as 'number of parts per hundred', interpret percentages and percentage changes as a fraction or a decimal, interpret these multiplicatively, express one quantity as a percentage of another, compare two quantities using percentages, and work with percentages greater than 100%.

Quick Search Ref: 10072

Correct: Correct.   **Wrong:** Incorrect, try again.   Open: Thank you.

Level 1: Understanding - Converting between fractions, decimals and percentages and expressing quantities as a fraction of another.

Required: 7/10   **Student Navigation:** on   **Randomised:** off

1. Express 42 as a percentage of 100.

a b c   ▪ `42` ▪ `42%`

2. Write 10 out of 20 as a percentage.

a b c   ▪ `50%` ▪ `50`

3. Express 3/8 as a decimal.

a b c   ▪ `0.375`

4. Express 0.375 as a percentage to one decimal place.

a b c   ▪ `37.5%` ▪ `37.5`

5. What calculation is used to convert from a fraction to a percentage?

☐ ☒ ☐
1/4

- ▪ **Divide the numerator by 100 and multiply by the denominator.**
- ▪ **Divide the numerator by the denominator and multiply by 100.**
- ▪ **Divide the denominator by 100 and multiply by the numerator.**
- ▪ **Divide the denominator by the numerator and multiply by 100.**

6. What is 18 out of 40 as a percentage?

☐ ☒ ☐
1/4

▪ `45%` ▪ `18/40` ▪ `0.45` ▪ `45`

7. What percentage of the grid is shaded?

a b c   ▪ `55%` ▪ `55`

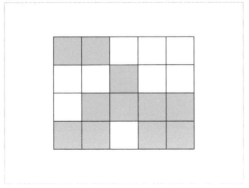

8. What is 110 as a percentage of 500?

a b c   ▪ `22%` ▪ `22`

9. Express 62 as a percentage of 80. Give your answer to **one decimal place**.

a b c   ▪ `77.5` ▪ `77.5%`

10. What percentage of the grid is shaded?

a b c   ▪ `40%` ▪ `40`

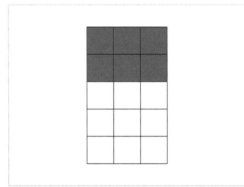

Level 2: Fluency - Expressing quantities as a fraction of another quantity, percentages greater than 100% (with and without calculator).

Required: 7/10   **Student Navigation:** on
Randomised: off

11. What is £80 out of £640 as a percentage? Give your answer to the **nearest whole percent**.

a b c   ▪ `13` ▪ `13%`

12. Natalie makes 350 keyrings and sells 280 at a village market. What percentage of the keyrings does she sell?

a b c   ▪ `80%` ▪ `80`

13. Express 45 as a percentage of 32. Give your answer to **one decimal place**.

a b c   ▪ `140.6` ▪ `140.6%`

Level 2: cont.

14. There are 32 children is class 8B. On Tuesday 4 of the children were absent.
What percentage of the children were absent on Tuesday?
Give your answer to **one decimal place**.

- 12.5 - 12.5%

15. Express 4.5 as a percentage of 9.65.
Give your answer to **two decimal places**.

- 46.63 - 46.63%

16. Jenny buys a laptop for £275 in a sale. Its original price was £385. How many percent is the laptop reduced by in the sale?
Give your answer to the **nearest whole percent**.

- 29% - 29

17. Express 129 grams as a percentage of 2.5 kilograms.
Give your answer to **two decimal places**.

- 5.16% - 5.16

18. Hamza buys a car for £900 and sells it for £1,280. Write the sale price as a percentage of the original price.
Give your answer to the **nearest whole percent**.

- 142% - 142

19. What is 87p as a percentage of £3?

- 29 - 29%

20. What is 300 as a percentage of 120?

- 250 - 250%

Level 3: Reasoning - Ordering, comparing and explaining quantities as percentages of other quantities.

Required: 4/6   **Student Navigation:** on
Randomised: off

21. Nathan scores 70% on a test. Select all of the scores which could be Nathan's.

3/6

- 32 out of 40 - 35 out of 50 - 41 out of 60
- 58 out of 70 - 56 out of 80 - 63 out of 90

22. Blake flipped a coin and recorded whether it landed on heads or tails. The ratio of heads to tails was 16:12. What percentage of the coin flips landed on heads?
Give your answer to the **nearest whole number**.

- 57 - 57%

23. Emma is calculating 640 as a percentage of 300 and gets 46.9% as her answer to one decimal place.
What mistake has Emma made? Explain how you know.

24. Select the symbol that makes the statement true.
56 out of 60 __ 78 out of 90

1/3

- < - = - >

25. Three friends played a computer game. Who got the highest percentage score?
Jason: 24/30.
Taylor: 35/40
1/3 Safia: 42/50.

- Jason - Taylor - Safia

26. Caleb scored 60% in a test. He says, "If I'd have got 10 more marks, I would have scored 70%".
Is Caleb correct? Explain your answer.

Level 4: Problem Solving - With quantities as percentages of other quantities.

Required: 6/6   **Student Navigation:** on
Randomised: off

27. Charlotte is spinning a number spinner. What percentage (to the nearest whole number) of the spins land on an odd number?
14 spins land on **1**.
12 spins land on **2**.
6 spins land on **3**.
16 spins land on **4**.
7 spins land on **5**.

- 49% - 49

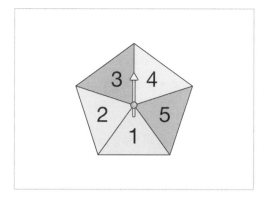

Level 4: cont.

28. Arrange Raj's subjects in order, starting with the lowest percentage score.

- French - Biology - English - Maths - History

Raj's test scores	
Maths	34 out of 50
French	18 out of 30
Biology	25 out of 40
History	17 out of 20
English	27 out of 40

29. Win and Jess make strawberry tarts to sell at their school fair.
Win makes 64 tarts and sells 48 of them.
Jess sells 82 of the 90 tarts that she makes.
To the nearest whole percent, what percentage of the tarts do the girls sell altogether?

- 84% - 84

30. The table shows the number of season tickets sold for four football teams. Which team has the highest percentage of child season ticket holders?

1/4
- Flamingo Warriors - Pelican Rovers - Macaw United
- Peacock Wanderers

team	adult	child	total
Flamingo Warriors	28,965		55,498
Pelican Rovers	21,450	21,391	
Macaw United		29,101	58,904
Peacock Wanderers	21,582		34,076

31. A board game has 84 counters. What percentage of the counters are **green**?
Give your answer to **one decimal place**.
12 counters are **blue**.
1/2 of the counters are **red**.
25% of the counters are **yellow**.
The remaining counters are **green**.

- 10.7 - 10.7%

32. To get from **start** to **finish** you can only pass through squares that convert to whole percentages. You can also only move left/right/up/down.
Arrange the colours to represent the path you take.

- Yellow - Pink - Green - Blue - Purple

start	72 out of 90	$\frac{350}{600}$	850 as a % of 1100
56 cm out of 94 cm	84 as a % of 210	£96 out of £120	$\frac{24}{70}$
$\frac{84}{78}$	640 g out of 1.2 kg	482 as a % of 964	650 as a % of 320
$\frac{64}{182}$	982 as a % of 548	$\frac{112}{280}$	finish

Find a percentage of an amount

Competency: Interpret fractions and percentages as operators

Quick Search Ref: 10067

Correct: Correct.   Wrong: Incorrect, try again.   Open: Thank you.

Level 1: Understanding - Simple percentages that can be calculated without a calculator.

❋ **Required:** 7/10   ❋ **Student Navigation:** on   ❋ **Randomised:** off

1. Give 10% as a fraction in its simplest form.

 ▪ 1/10

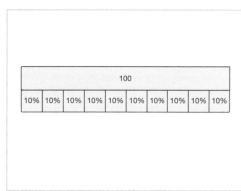

2. What is 10% of 48?

 ▪ 4.8

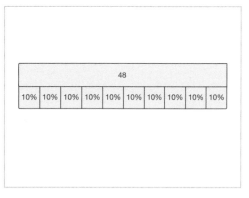

3. What is 1% as a fraction?

 ▪ 1/100

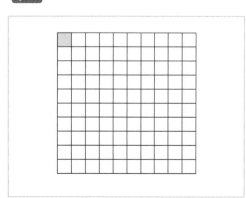

4. What is 1% of 62?

 ▪ 0.62

5. Calculate 11% of 40.

 ▪ 4.4

6. Give 62% as a decimal.

 ▪ 0.62

7. What is 7% as a decimal?

 ▪ 0.07

8. Find 12% of 60.

 ▪ 7.2

9. Calculate 21% of 210.

 ▪ 44.1

10. What is 6% as a decimal?

 ▪ 0.06

Level 2: Fluency - Calculating percentages of an amount using a calculator.

❋ **Required:** 7/10   ❋ **Student Navigation:** on
❋ **Randomised:** off

11. What is 62% of 48?

 ▪ 29.76

12. Calculate 7% of 98.

 ▪ 6.86

13. Find 128% of 64.

 ▪ 81.92

Level 2: *cont.*

14. The average cat spends 66% of its life asleep.
If a cat lives for 15 years, how long does it spend sleeping?
Give your answer to the nearest number of years.

- 10

15. 11% of the population is left-handed.
In a school with 1,200 students, how many of them would you expect to be **right-handed**?

- 1068

16. Cucumbers are 96% water.
If a cucumber weighs 220 grams, how much of it is water?
Give your answer to the nearest gram.

- 211

17. On average, 56% of typing is completed using your left hand.
If you type 689 characters, how many of these would you expect to type with your **right hand**?
Give your answer to the nearest whole number.

- 303

18. 61% of British people have brown hair.
How many children in a class of 30 would you expect to have brown hair?

- 18

19. A test is marked out of 40 and Charlie gets a score of 37%.
How many **marks** does Charlie get?
Give your answer to the nearest whole number.

- 15

20. A team collaborated on a 1,200 word report.
Rebecca wrote 36% of the report.
How many words did she write?

- 432

Level 3: Reasoning - Showing an understanding of the concepts in finding a percentage of a quantity.

- **Required:** 5/5   - **Student Navigation:** on
- **Randomised:** off

21. Alex says that 10% is the same as 1/10, so 20% is the same as 1/20.
Explain why she is wrong.

22. Is it true that 60% will always be a larger amount than 20%?

- Always true - Sometimes true - Never true

1/3

23. "25% of 24 is the same as 24% of 25".
Select **true or false** and the **reason why**.

- True - False

2/5
- Because the numbers are very closer to each other so you'll get the same answer.
- The percentages are different so the answers can't be the same.
- 0.24 x 25 is the same as 0.25 x 24.

24. A toy is reduced by 20% in a sale.
The next week, the new price is reduced by 10%.
By how many percent has the toy been reduced altogether?

- 28

25. "If the price of a train ticket is increased by 20%, and then the following week the new price is reduced by 20%, the ticket will be back to its original price".

2/4
Select **true or false** and the **reason why**.

- True - False
- You are taking 20% off the price and then adding it back on.
- You are finding 20% of two different numbers, so the final price will be different to the original price.

Level 4: Problem Solving - Finding fractions of amount to solve complex problems in real-life contexts.

- **Required:** 5/5   - **Student Navigation:** on
- **Randomised:** off

26. In a sale everything is reduced by 70%. Charlotte spends £48.
How much money does she save in pounds?
Include the £ sign in your answer.

- £112

27. Three friends win the lottery.
Finn gets 25%, Aiden gets 30% and Ethan gets 45%.
Ethan gets £45,000 more than Aiden.
How much does Finn get in pounds?

- 75000

28. How many letters does the postwoman deliver in total?
Monday: 800 letters.
Tuesday: 30% more than Monday.
Wednesday: 20% less than Tuesday.
Thursday: 25% more than Wednesday.
Friday: 40% less than Thursday.

- 4336

29. Asher is renovating his house.

In the first year he completes 25% of the work.

In the second year he completes 30% of the work that's left.

What percentage of work does he still need to do?

Include the % symbol in your answer.

- 52.5%

30. In a school, 55% of the students are girls.

48% of the girls bring a packed lunch.

24% of the boys bring a packed lunch.

What percentage of pupils in the school bring a packed lunch?

Give your answer to one decimal place.

Include the % symbol in your answer.

- 37.2%

Increase an amount by a percentage

Competency: Solve problems involving percentage change, including: percentage increase, decrease and simple interest in financial mathematics.

Quick Search Ref: 10018

Correct: Correct.   Wrong: Incorrect, try again.   Open: Thank you.

Level 1: Understanding - How to increase an amount by a percentage.

✿ Required: 7/10   ✿ Student Navigation: on   ✿ Randomised: off

1. If you increase an amount by 20%, how much will you have as a percentage of the original amount?

   a b c ▪ 120 ▪ 120%

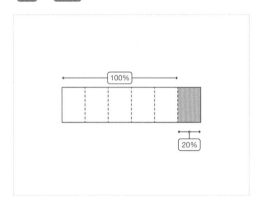

2. If you increase an amount by 5%, how much will you have as a percentage of the original amount?

   a b c ▪ 105 ▪ 105%

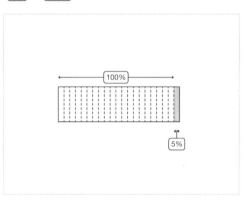

3. Express 167% as a decimal.

   1 2 3 ▪ 1.67

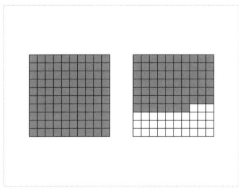

4. Select the calculation you would use to find 12% of an amount.

   ☐ ☒ ☐ ▪ x 0.12 ▪ x 1.12 ▪ x 12 ▪ x 112

   1/4

5. What would you multiply an amount by to **increase** by 5%?

   1 2 3 ▪ 1.05

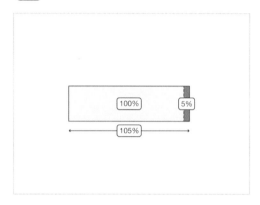

6. Select the two calculations you could use to **increase** 60 by 34%.

   ☐ ☒ ☐ ▪ 60 + (60 x 0.34) ▪ 60 x 0.34 ▪ 60 x 1.34

   2/3

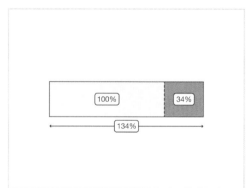

7. Select the two calculations you would use to **increase** 35 by 7%?

■ 35 + (35 x 0.07)   ■ 35 x 0.07   ■ 35 x 0.7   ■ 35 x 1.07

2/4

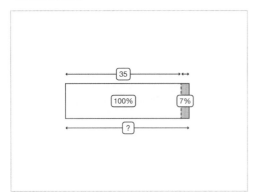

8. Select the calculation you could use to find 3% of an amount.

■ x 0.03   ■ x 3   ■ x 0.3

1/3

9. What would you multiply an amount by to **increase** by 73%?

■ 1.73

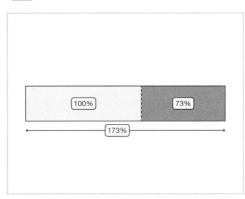

10. Select the two calculations you could use to **increase** 94 by 62%?

■ 94 x 0.62   ■ 94 x 1.62   ■ 94 + (94 x 0.62)

2/3

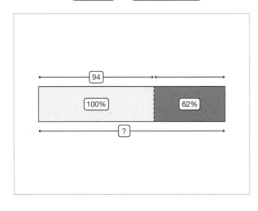

Level 2: Fluency - Calculate a percentage increase including worded problems.

✿ **Required:** 7/10   ✿ **Student Navigation:** on
✿ **Randomised:** off

11. Find the new amount when 54 is increased by 32%.

■ 71.28

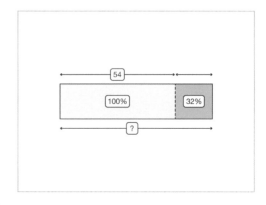

12. What is 21% more than 84?

■ 101.64

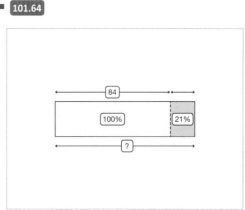

13. What is 65% more than 28?

■ 46.2

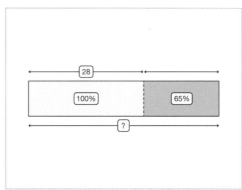

14. A restaurant adds a 12% service charge
a
b to every bill. What is the new total after the
c service charge is added to a bill of £48?
Give your answer in pounds and pence.
Include the £ sign in your answer.

▪ £53.76

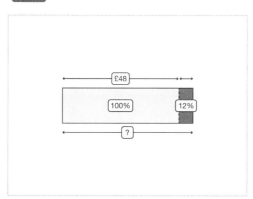

15. Charlotte is training to run a marathon.
a
b She completes a 14 mile run in week 5 of training
c and wants to increase this by 7% in week 6. How
far will her run be in week 6?
Give your answer to the nearest mile.
Include the units m (miles) in your answer.

▪ 15 m ▪ 15 miles

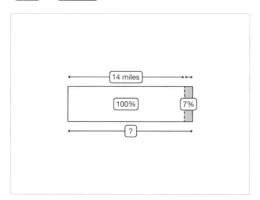

16. Lily makes earrings. A pair of earrings costs £2.80
a
b to make and she wants to earn 150% profit. How
c much should she sell them for?
Give your answer in pounds and pence.
Include the £ sign in your answer.

▪ £7 ▪ £7.00

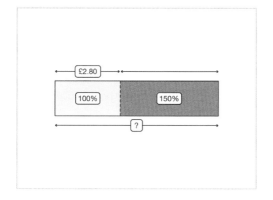

17. In London, there was 4% more rainfall in January
a
b than in May.
c If there were 50 millimetres of rainfall in May, how
many millimetres of rainfall were there in January?
Include the units mm (millimetres) in your answer.

▪ 52 mm ▪ 52 millimetres

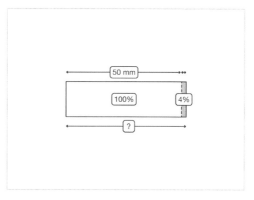

18. What is 42% more than 125?

1
2 ▪ 177.5
3

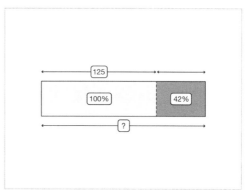

19. Find the new amount when 87 is increased by
1
2 57%.
3
▪ 136.59

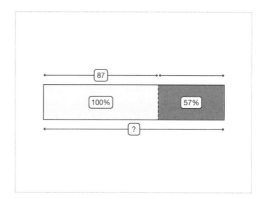

20. 20% VAT is added to the cost of every laptop sold.
The price of the laptop before VAT is £349. How
much will it cost with VAT added?
Give your answer in pounds and pence.
Include the £ sign in your answer.

a
b
c

- £418.80

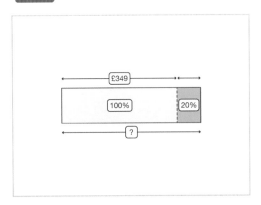

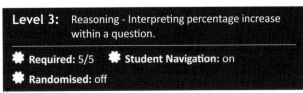

Required: 5/5   **Student Navigation:** on
Randomised: off

21. If an amount is doubled, what percentage has it
been **increased by**?

a
b
c

- 100 - 100%

22. If an amount has been increased by 100%, what
do you need to **multiply** the new value by to get
the original amount?

a
b
c

- 1/2 - 0.5

23. What do you need to multiply by if you want to
increase an amount to 150% of its original value.

1
2
3

- 1.5

24. Train fares are increased by 10%. If a ticket now
costs £13.20, what was its original price?
Include the £ sign in your answer.

a
b
c

- £12 - £12.00

25. A number increases by 50% each year.
By how many percent will it have increased after 3
years?

a
b
c

- 237.5% - 237.5

Required: 5/5   **Student Navigation:** on
Randomised: off

26. Mr Williams has a pond that is 12 metres wide
and 36 metres long.
After three months he reduces the width of
the pond to 8 metres.
By what percentage must he increase the length to
maintain the same area as before?

a
b
c

- 50 - 50%

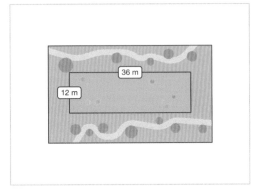

27. A strain of bacteria increases its population as
follows:
20% in the first hour.
30% in the second hour.
50% in the third hour.
What is the overall percentage increase in the first
3 hours?

a
b
c

- 134 - 134%

28. The cardboard box has a volume of 3,600 cm³.
Side a is increased by 32%.
Side b is increased by 2%.
Side c is increased by 19%.
What is the volume of the new box?
Give your answer to the nearest cubic centimetre.
Don't include the units in your answer.

a
b
c

- 5768 - 5,768

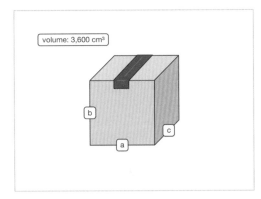

29. Liam has money in his savings account.

He receives 1.5% interest at the end of the first year.
At the end of the second year he gets 2.5% interest.
If he now has £208.08, how much did he start with?

▪ £200 ▪ £200.00

30. Hannah makes 140% profit on her cakes, which she sells for £4.80.

If she wants to increase the price so that she is earning a 200% profit. What is the new price of a cake?
Include the £ sign in your answer.

▪ £6.00 ▪ £6

Compare and convert between percentages, fractions and decimals

Competency: Define percentage as 'number of parts per hundred', interpret percentages and percentage changes as a fraction or a decimal, express one quantity as a percentage of another, compare two quantities using percentages, and work with percentages > 100%.

Quick Search Ref: 10049

Correct: Correct.　　Wrong: Incorrect, try again.　　Open: Thank you.

Level 1: Understanding - Convert between fractions, decimals and percentages.

✱ **Required:** 10/10　　✱ **Student Navigation:** on　　✱ **Randomised:** off

1.　What is 37% written as a **decimal**?

 ▪ 0.37

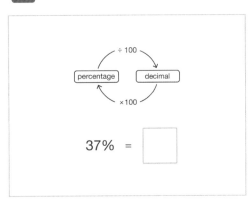

2.　Give 0.25 as a **percentage**.
Include the % sign in your answer.

▪ 25%

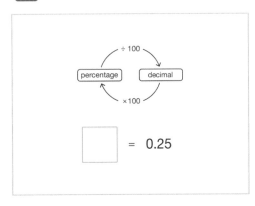

3.　What is 60% written as a **fraction** in its simplest form?

 ▪ 3/5

$$60\% = \frac{?}{100} = \frac{?}{5}$$

4.　Convert 0.54 to a **fraction** in its simplest form.

▪ 27/50

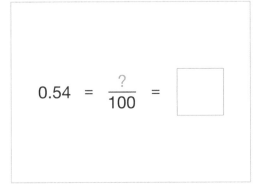

$$0.54 = \frac{?}{100} = \boxed{}$$

5.　What is 7/8 as a **decimal**?

▪ 0.875

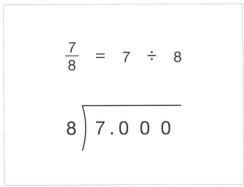

$$\frac{7}{8} = 7 \div 8$$

$$8\overline{)7.0\,0\,0}$$

6.　What is 9/25 as a **decimal**?

▪ 0.36

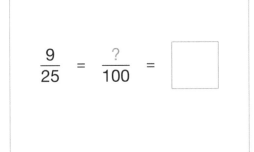

$$\frac{9}{25} = \frac{?}{100} = \boxed{}$$

Level 1: *cont.*

7. Convert 2/7 to a **percentage**.

Round your answer to the nearest whole percent.
Include the % sign in your answer.

▪ 29%

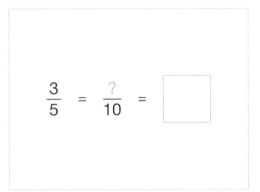

$$\frac{2}{7} = 2 \div 7$$

$$7\overline{)2.000}$$

8. What is 3/5 as a **decimal**?

▪ 0.6

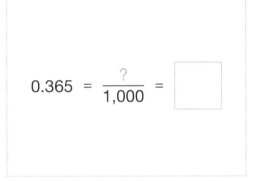

$$\frac{3}{5} = \frac{?}{10} = \boxed{}$$

9. Give 0.365 as a **fraction** in its simplest form.

▪ 73/200

$$0.365 = \frac{?}{1,000} = \boxed{}$$

10. Give 7/20 as a **percentage**.

Include the % sign in your answer.

▪ 35%

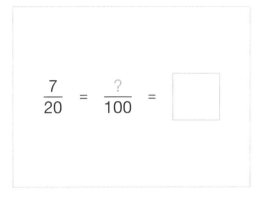

$$\frac{7}{20} = \frac{?}{100} = \boxed{}$$

Level 2: Fluency - Questions in context, comparing two amounts more than 100%.

✿ **Required:** 7/10   ✿ **Student Navigation:** on
✿ **Randomised:** off

11. The Earth's atmosphere consists of 78% nitrogen.
What is this amount as a fraction in its simplest form?

▪ 39/50

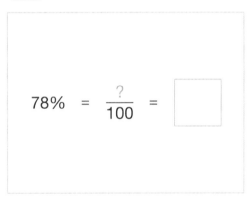

$$78\% = \frac{?}{100} = \boxed{}$$

12. 5 out of every 8 students in a school are boys.
What is this as a percentage to one decimal place?
Include the % sign in your answer.

▪ 62.5%

13. What missing symbol makes the following statement correct?
0.37 ___ 3/8

1/3 ▪ < ▪ > ▪ =

$$0.37 \ \boxed{} \ \frac{3}{8}$$

14. If approximately 60% of the adult human body is water, what fraction is not water?
Give your answer in its simplest form.

a b c

▪ 2/5

15. Donna adds 400 ml (millilitres) of water to 100 ml of cordial.
What percentage of Donna's drink is cordial?
Include the % sign in your answer.

a b c

▪ 20%

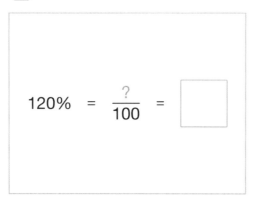

400 ml water

100 ml cordial

16. What is 120% as a decimal?

1 2 3

▪ 1.2

$$120\% = \frac{?}{100} = \boxed{}$$

17. Spaghetti contains 0.32 g of carbohydrates per gram of spaghetti. What fraction of spaghetti is made up of carbohydrates?
Give your answer in its simplest form.

a b c

▪ 8/25

18. The atmosphere of Venus is 96% carbon dioxide. The atmosphere of Mars is 19/20 carbon dioxide. Which planet's atmosphere has the highest percentage of carbon dioxide?

1/2

▪ Mars   ▪ Venus

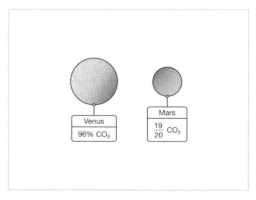

Venus
96% CO_2

Mars
$\frac{19}{20}$ CO_2

19. Javed scored 48 out of 75 in his English test and got 65% in his Maths test. Which subject did Javed do better in?

1/2 ▪ Maths ▪ English

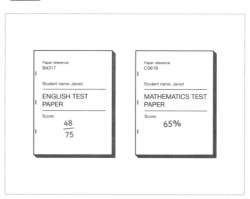

Paper reference:
B4317

Student name: Javed

ENGLISH TEST PAPER

Score:
$\frac{48}{75}$

Paper reference:
C5619

Student name: Javed

MATHEMATICS TEST PAPER

Score:
65%

20. Tony buys a second-hand skateboard for £25. He refurbishes it and sells it for £40.
What is the sale price as a percentage of the price Tony paid for the skateboard?
Include the % sign in your answer.

a b c

▪ 160%

Level 3: Reasoning - Misconceptions when converting and comparing fractions, decimals and percentages.

✿ **Required:** 5/5   ✿ **Student Navigation:** on
✿ **Randomised:** off

21. Katie says '3 is less than 25 so 0.3 is less than 0.25'.
Is Katie right? Explain your answer.

a
b
c

22. What missing numerator makes this statement correct?
1
2
3
$0.7 < \_\_/5 < 85\%$

▪ 4

$$0.7 \quad < \quad \frac{?}{5} \quad < \quad 85\%$$

23. Iain has completed his maths homework in red pen. One answer is incorrect. What should Iain have written instead of this answer?

a
b
c

▪ 40   ▪ 40%

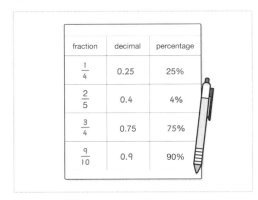

fraction	decimal	percentage
$\frac{1}{4}$	0.25	25%
$\frac{2}{5}$	0.4	4%
$\frac{3}{4}$	0.75	75%
$\frac{9}{10}$	0.9	90%

24. What **fraction** comes next in the following sequence?
a
b
c
1/20, 0.1, 15%, 1/5, 0.25, 30%, ___.

▪ 7/20

$\frac{1}{20}$ 0.1   15% $\frac{1}{5}$ 0.25   30%   ☐

25. Camila has sorted the following amounts into ascending order.
a
b
c
What mistake has Camila made?
45%, 0.4, 1/2, 0.48.

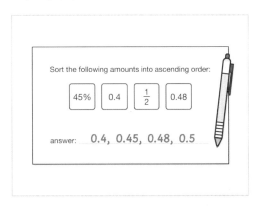

Sort the following amounts into ascending order:

45%   0.4 $\frac{1}{2}$ 0.48

answer: 0.4, 0.45, 0.48, 0.5

Level 4: Problem solving - Ordering fractions, decimals and percentages.

✿ **Required:** 5/5   ✿ **Student Navigation:** on
✿ **Randomised:** off

26. Sort the amounts in ascending order (lowest value) first.
↑
↓

▪ 0.4   ▪ 3/7   ▪ 4/9   ▪ 45%

Level 4: cont.

27. The table shows the protein content per gram of various foods.
List the foods by protein content in descending order (highest value first).

- Low fat cheddar 1/3 ∎ Chicken breast 8/25
- Cheddar cheese 1/4 ∎ Tuna 23.5 % ∎ Almonds 0.211

food	protein per g
almonds	0.211
cheddar cheese	$\frac{1}{4}$
chicken breast	$\frac{8}{25}$
low fat cheddar	$\frac{1}{3}$
tuna	23.5%

28. The table shows Dominic's exam results. Arrange the subjects in order, starting with the best result.

- Physics 28/40 ∎ French 68% ∎ Maths 13/20
- English 16/25 ∎ Art 63%

exam results	
art	63%
English	$\frac{16}{25}$
French	68%
maths	$\frac{13}{20}$
physics	$\frac{28}{40}$

29. Sort the amounts in descending order, highest first.

- 2.3 ∎ 9/4 ∎ 220% ∎ 2.18 ∎ 13/6

30. The table shows fifteen amounts. Fourteen of the amounts can be made into seven equivalent pairs. Which is the odd one out?

- 60%

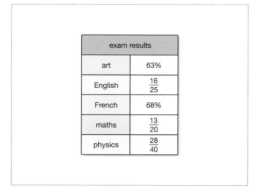

$\frac{5}{8}$	55%	$\frac{33}{40}$	0.5	36%
0.8	$\frac{9}{25}$	0.625	$\frac{11}{20}$	$\frac{4}{5}$
$\frac{1}{2}$	60%	$\frac{7}{20}$	$82\frac{1}{2}$%	0.35

Mathematics Y8

Ratio and Proportion

Ratio

Proportion

Divide a Quantity into a Given Ratio

Competency: Divide a given quantity into two parts in a given part: part or part: whole ratio; express the division of a quantity into two parts as a ratio.

Quick Search Ref: 10316

Correct: Correct.   **Wrong:** Incorrect, try again.   **Open:** Thank you.

Level 1: Understanding - Dividing an amount in a ratio supported by bar modelling.

✱ **Required:** 7/10   ✱ **Student Navigation:** on   ✱ **Randomised:** off

1. Courtney and Drew share £24 in the ratio 1:2. How much does Courtney receive?
Include the £ sign in your answer.

■ £8 ■ £8.00

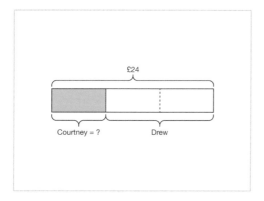

2. Alan and Brian share 48 football stickers in the ratio 1:7. How many stickers does Alan receive?

■ 6

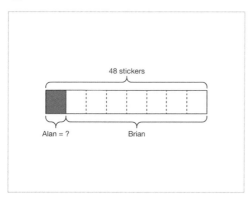

3. Kyle and Lucy share 84 counters in the ratio 6:1. How many counters does Lucy receive?

■ 12

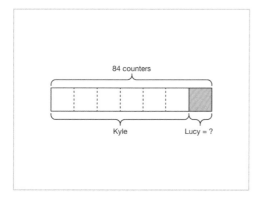

4. Robyn and Sara share £60 in the ratio 2:3. How much does Robyn receive?
Include the £ sign in your answer.

■ £24.00 ■ £24

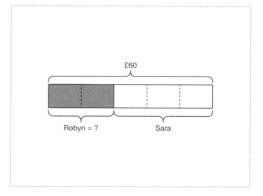

5. The ratio of boys to girls in a class is 5:3. If there are 32 students in the class, how many boys are there?

■ 20

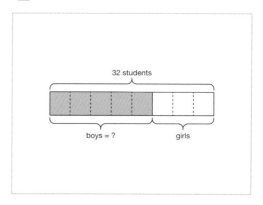

6. Mike and Nadia share £70 in the ratio 5:2. Select the amounts they receive.
2 answers required.

■ £10 ■ £60 ■ £50 ■ £20 ■ £40 ■ £30

2/6

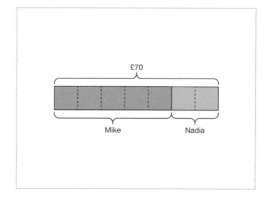

7. The ratio of adults to children in a tennis club is 7:4. If the club has 154 members altogether, how many of them are children?

- 56

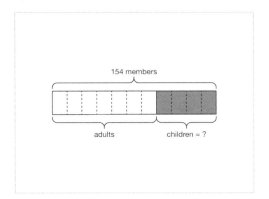

8. Georgia and Hamza share 52 marbles in the ratio 9:4. How many marbles does Hamza get?

- 16

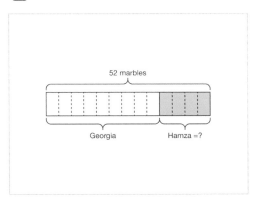

9. Ruben and Sean share 180 crayons in the ratio 5:4. How many crayons do they each receive?

- 120 and 60 ▪ 140 and 40 ▪ 110 and 70 ▪ **100 and 80**
1/5 ▪ 130 and 50

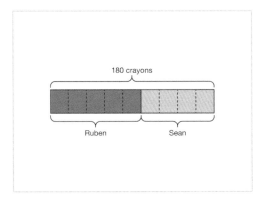

10. Melissa and Nasim share their newspaper deliveries in the ratio 7:3. If there are 120 newspapers altogether, how many does Melissa deliver?

- 84

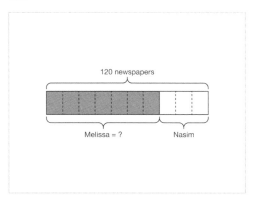

Level 2:   Fluency - Three-part ratios and working with larger numbers.

✹ **Required:** 7/10   ✹ **Student Navigation:** on
✹ **Randomised:** off

11. In a race, the ratio of male to female runners is 5:3. If there are 1,360 competitors, how many are female?

- 510

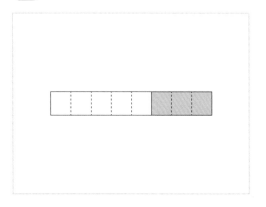

12. Aisha, Beth and Charlie share 135 building blocks in the ratio 1:3:5. How many blocks does Charlie receive?

- 75

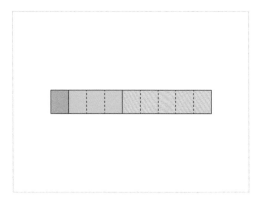

Level 2: *cont.*

13. Owen and Paula divide 126 coins in the ratio 7:2.
How many more coins does Owen have than
Paula?

- 70

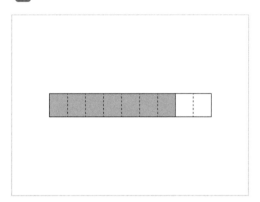

14. A triangle has angles in the ratio 2:3:5. How many
degrees is the smallest angle?
Don't include the units in your answer.

- 36

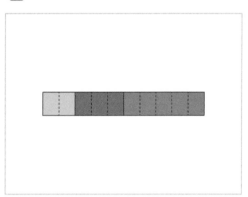

15. A school has male and female students in the ratio
4:3. If there are 455 students in total, how many
male students are there?

- 260

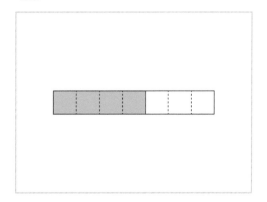

16. Jasmin and Kylie share 136 pens in the ratio 6:11.
How many more pens does Kylie receive than
Jasmin?

- 40

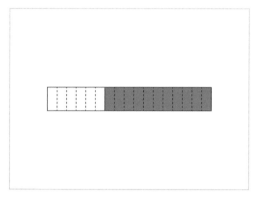

17. A rectangle has sides in the ratio 5:7. If the
perimeter of the rectangle is 216 centimetres,
what is the length of the longest side?
Include the units centimetres (cm) in your answer.

- 63 cm - 63 centimetres

18. A box has red, white and blue counters in the ratio
3:5:7. If there are 1,050 counters in total, how
many are blue?

- 490

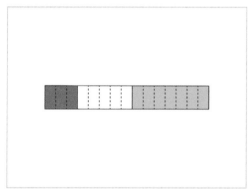

19. An orange paint is made by mixing yellow paint
and red paint in the ratio 6:1. If Chloe needs 91
litres of orange paint, how many litres of yellow
paint does she need?
Include the units litres (l) in your answer.

- 78 litres - 78 l

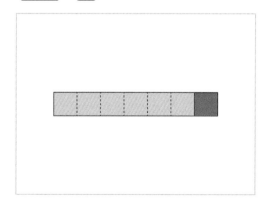

Level 2: *cont*.

20. The ratio of women to men who are over 90 in the UK is 7:3. How many women would you expect there to be in a city with 2,320 people over the age of 90?

a b c

▪ 1624 ▪ 1,624

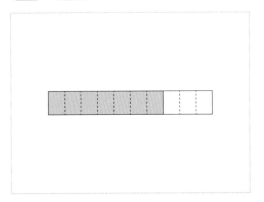

Level 3: Reasoning - Comparing ratios and misconceptions.

❋ **Required:** 5/5   ❋ **Student Navigation:** on
❋ **Randomised:** off

21. Daisy and Ella are dividing 1,080 marbles in the following ratios. Which two ratios will give Ella the same amount of marbles?

2/7 ▪ 4:1 ▪ 3:3 ▪ 15:10 ▪ 2:3 ▪ 5:7 ▪ 3:2 ▪ 10:2

22. A pastry recipe uses water, butter and flour mixed in the ratio 2:5:10.
If Ryan has 200 grams of butter, what is the maximum mass of the pastry he can make?
Include the units grams (g) in your answer.

a b c

▪ 680 grams ▪ 680 g

23. Maria and Nisa are sharing stickers in the ratio 1:3. Maria says she receives one third of the stickers. Is she correct? Explain your answer.

a b c

1:3 means I get $\frac{1}{3}$

24. Rhys and Safia are dividing an amount of money in the given ratios. Which ratio will give Rhys the most amount of money?

☐☒☐

1/5 ▪ 3:4 ▪ 2:3 ▪ 11:15 ▪ 3:5 ▪ 13:20

25. Concrete is made by mixing cement, sand and gravel in the ratio 1:3:4.
Dave wants to make 240 kilograms (kg) of concrete and has 35 kg of cement, 85 kg of sand and 130 kg of gravel.
Does Dave have enough cement, sand and gravel to make the concrete mix? Explain your answer.

a b c

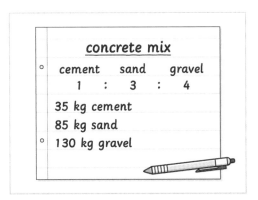

concrete mix

cement		sand		gravel
1	:	3	:	4

35 kg cement
85 kg sand
130 kg gravel

Level 4: Problem Solving - Multiple ratios and changes to ratios.

❋ **Required:** 5/5   ❋ **Student Navigation:** on
❋ **Randomised:** off

26. Sarah and Tony buy a painting for £240. Sarah pays £90 and Tony pays £150.
A year later they sell the painting for £560. How much of the money should Tony receive?
Include the £ sign in your answer.

a b c

▪ £350 ▪ £350.00

Level 4: *cont.*

27. A 20 centimetre piece of string is used to make a
a square. The same piece of string is then used to
b make a rectangle with sides in the ratio 2:3.
c What is the ratio of the area of the rectangle to
the area of the square?

▪ 24:25

area of rectangle   area of square
?   :   ?

28. An isosceles triangle has two sides in the ratio 2:5.
a If the perimeter of the triangle if 36 centimetres,
b what is the length of the longest side of the
c triangle in centimetres?
Include the units centimetres (cm) in your answer.

▪ 15 centimetres ▪ 15 cm

29. The grid shows the number of counters in nine
1 different bags and the ratio of blue counters to red
2 counters.
3 Four pairs of bags have the same number of blue
counters. How many blue counters are in the
remaining bag?

▪ 28

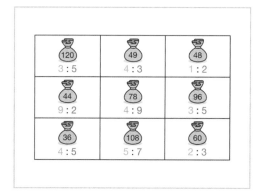

120	49	48
3 : 5	4 : 3	1 : 2
44	78	96
9 : 2	4 : 9	3 : 5
36	108	60
4 : 5	5 : 7	2 : 3

30. Justin and Kyra have ages in the ratio 1:2. In 12
1 years time their ages will be in the ratio 4:5. How
2 old is Kyra now?
3

▪ 8

Find Missing Amounts in a Ratio

Competency: Divide a given quantity into two parts in a given part: part or part: whole ratio; express the division of a quantity into two parts as a ratio.

Quick Search Ref: 10314

Correct: Correct.   Wrong: Incorrect, try again.   Open: Thank you.

Level 1:   Understanding - Finding missing parts in a ratio supported by bar modelling.

✿ **Required:** 7/10   ✿ **Student Navigation:** on   ✿ **Randomised:** off

1. Jack and Owen have marbles in the ratio 1:2.
If Jack has 8 marbles, how many marbles does Owen have?

▪ 16

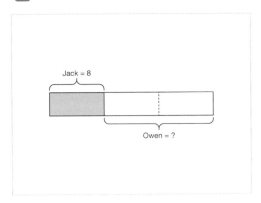

2. The ratio of pencils to pens in a box is 1 to 5.
If there are 4 pencils, how many pens are there?

▪ 20

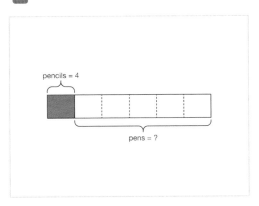

3. A class has 2 boys for every 3 girls.
The class has 12 boys, how many girls are there?

▪ 18

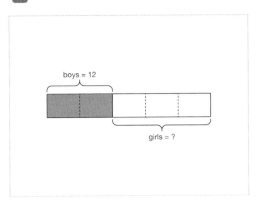

4. A bag has green and yellow counters in the ratio 3:1. If there are 12 yellow counters, how many green counters are there?

▪ 36

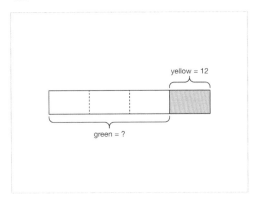

5. A laptop screen has a height and width in the ratio 2:3.
How wide is the screen if it is 24 centimetres high?
Include the units centimetres (cm) in your answer.

▪ 36 centimetres ▪ 36 cm

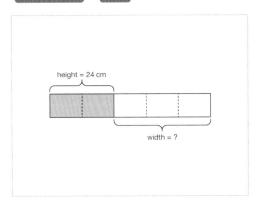

6. The ratio of tables to chairs in a restaurant is 2:7.
If there are 28 tables, how many chairs are there?

▪ 98

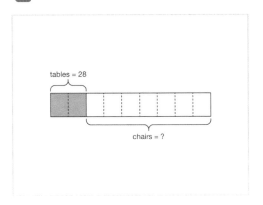

Level 1: *cont.*

7. Jenny and Marcello share some money in the ratio
   a 3:5.
   b If Marcello receives £30, how much does Jenny
   c receive?
   Include the £ sign in your answer.

   ▪ £18.00 ▪ £18

   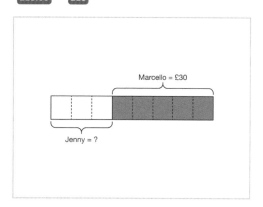

8. A school trip has Year 7 and Year 8 students in the
   1 ratio 4 to 5.
   2 If there are 60 Year 7 students, how many Year 8
   3 students are on the trip?

   ▪ 75

   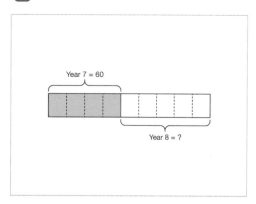

9. A mobile phone screen has a height and a width in
   a the ratio 3:2.
   b If the screen is 8 centimetres wide, how high is the
   c screen?
   Include the units centimetres (cm) in your answer.

   ▪ 12 cm ▪ 12 centimetres

   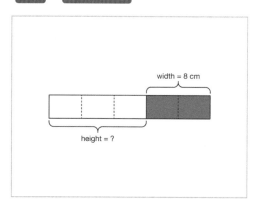

10. The ratio of apples to oranges in a shop is 4:9.
    1 If the shop has 72 apples, how many oranges are
    2 there?
    3

    ▪ 162

    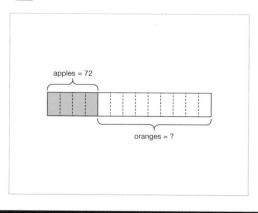

Level 2: Fluency - Finding missing parts in ratios with
larger numbers, converting units and finding
the total amount.

✹ **Required:** 7/10   ✹ **Student Navigation:** on
✹ **Randomised:** off

11. A race has male and female runners in the ratio
    1 4:3.
    2 If there are 216 males, how many females are in
    3 the race?

    ▪ 162

    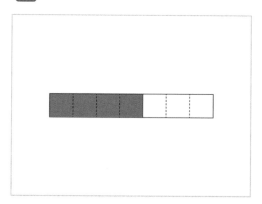

12. A bag contains black and white counters in the
    1 ratio 3:4.
    2 If the bag contains 24 black counters, how many
    3 counters are there in total?

    ▪ 56

    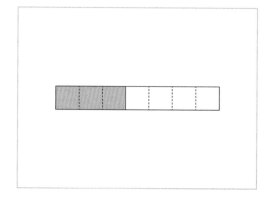

13. A builder makes mortar by mixing sand and
a cement in the ratio 5:1.
b If he has 2 kilograms of sand, how many grams of
c cement does he need?
Include the units grams (g) in your answer.

▪ 400 grams ▪ 400 g

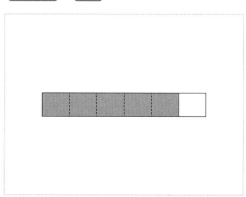

14. Purple paint is made by mixing 3 litres of blue
a paint with 5 litres of red paint.
b If Andy has 60 litres of blue paint, what is the
c maximum amount of purple paint he can make?
Include the units litres (l) in your answer.

▪ 160 l ▪ 160 litres

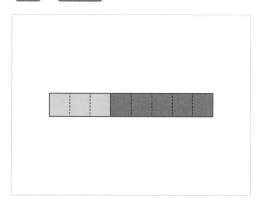

15. A ball pool has blue, green and yellow balls in the
1 ratio 2:3:5.
2 If there are 280 blue balls, how many yellow balls
3 are there?

▪ 700

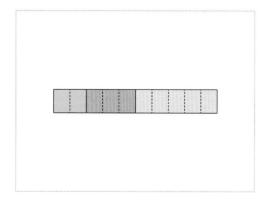

16. A farmer keeps cows and sheep in the ratio 7:9.
1 If he has 315 sheep, how many cows does he
2 have?
3

▪ 245

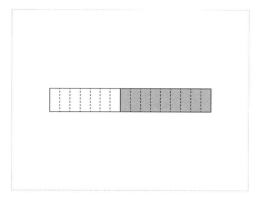

17. A triangle has side lengths in the ratio 2:6:7.
a If the length of the shortest side is 24 centimetres,
b what is the perimeter of the triangle?
c *Include the units centimetres (cm) in your answer.*

▪ 180 centimetres ▪ 180 cm

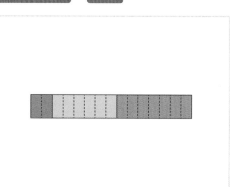

18. In a shipping yard the ratio of full containers to
1 empty containers is 5 to 2.
2 If the yard has 260 empty containers, how many
3 containers are there in total?

▪ 910

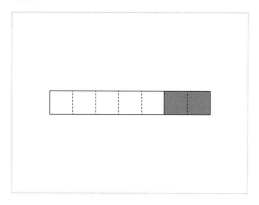

Level 2: *cont.*

19. A rectangle has sides in the ratio 5:7. If the shortest side is 20 centimetres, what is the perimeter of the rectangle?
Include the units centimetres (cm) in your answer.

a
b
c

▪ `96 centimetres` ▪ `96 cm`

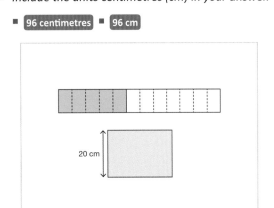

20. A bus has adult and child passengers in the ratio 9 to 4.
If there are 36 adults, how many passengers does the bus have?

1
2
3

▪ `52`

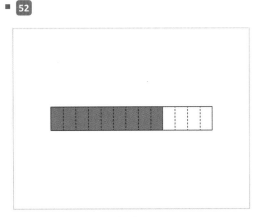

Level 3: Reasoning - Misconceptions and proportional reasoning.

✱ **Required:** 5/5   ✱ **Student Navigation:** on
✱ **Randomised:** off

21. Three orange paints are made by mixing red and yellow paint in the ratios of 1:2, 2:3 and 3:7
Which paint is the lightest shade?

☐
☒
☐

1/3   ▪ 1:2 ▪ 2:3 ▪ `3:7`

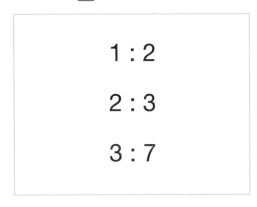

22. A company employs men and women in the ratio 5:4. If two men leave the company the ratio of men to women will be 3:4. Is this correct? Explain your answer.

a
b
c

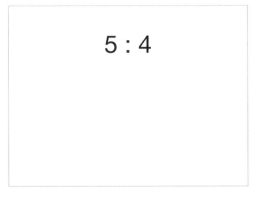

23. Two squares have side lengths in the ratio 1:2. What is the ratio of the area of the smaller square to the area of the larger square?

a
b
c

▪ `1:4`

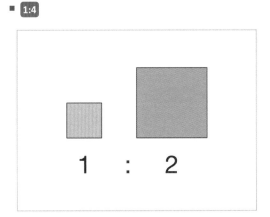

24. Jayne, Karina and Mary have beads in the ratio 3:5:8. If Karina has 18 more beads than Jayne, how many beads does Mary have?

1
2
3

▪ `72`

Jayne

Karina

Mary

Level 3: cont.

25. Darcey and Eva have ages in the ratio 1:3. In 12
a years time their ages will be in the ratio 2:3.
b Eva says in another 12 years time their ages will be
c in the ratio 3:3. Is Eva correct? Explain your
answer.

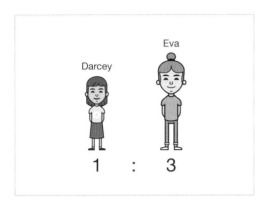

Level 4: Problem Solving - Changes in ratio and multi-
step problems.

✿ **Required:** 5/5 ✿ **Student Navigation:** on
✿ **Randomised:** off

26. A map is drawn with a scale of 1:25,000. If a walk
a is 24 centimetres on the map, how many
b kilometres will the actual walk be?
c *Include the units kilometres (km) in your answer.*

▪ 6 km ▪ 6 kilometres

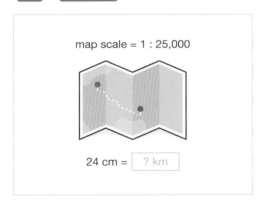

27. A bag contains red and blue counters in the ratio
1 1:3.
2 Teagan adds 6 red counters to the bag and the
3 ratio of red to blue counters is now 1:2. How many
blue counters are in the bag?

▪ 36

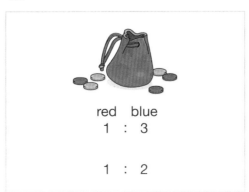

28. At a concert the ratio of adults to children is 5:2.
a Amongst the children, the ratio of boys to girls is
b 3:7. If there are 14,000 adults at the concert, how
c many boys are there?

▪ 1,680 ▪ 1680

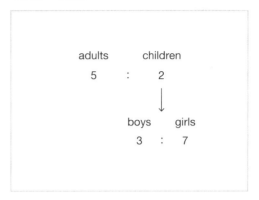

29. Roman and Paul are brothers with ages in the ratio
1 1:2.
2 In 10 years time their ages will be in the ratio 3:4.
3 How many years old is Paul now?

▪ 10

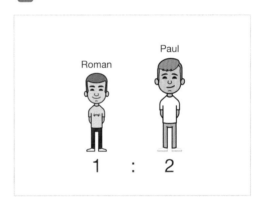

30. A box contains green and yellow marbles in the
1 ratio 3:5. Charlotte wants the ratio of green
2 marbles to yellow marbles to be 1:2. If there are
3 48 green marbles in the box, how many green
marbles must she remove?

▪ 8

Use Scale Diagrams and Maps

Competency: Use scale factors, scale diagrams and maps.

Quick Search Ref: 10380

Correct: Correct.   Wrong: Incorrect, try again.   Open: Thank you.

Level 1:   Understanding - Finding actual lengths using scales.

✱ **Required:** 7/10   ✱ **Student Navigation:** on   ✱ **Randomised:** off

1. A scale on a map tells you:

1/4

- ▪ the actual distance in kilometres from one town to another.
- ▪ what different symbols on the map mean.
- ▪ the orientation of the map.
- ▪ **the ratio between distances on the map and actual distances.**

2. A scale on a drawing, plan or map can be given in words or represented by a ratio.
A plan has a scale of 1 centimetre represents 2 metres.

1/6 What is the ratio of measurements on the plan to real life measurements?

- ▪ 1:2 ▪ 200:1 ▪ 2:100 ▪ 2:1 ▪ **1:200** ▪ 100:2

3. Reagan draws a television with a scale of 1 centimetre on the drawing representing 5 centimetres on the television.
a b c What is the actual height of the television?
Include the units cm (centimetres) in your answer.

- ▪ **50 centimetres** ▪ **50 cm**

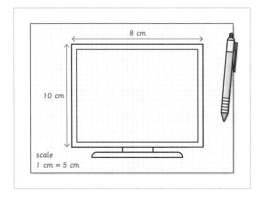

4. A floor plan of a classroom is drawn with a scale of 1 centimetre to 2 metres.
a b c What is the actual length of the classroom?
Include the units m (metres) in your answer.

- ▪ **16 m** ▪ **16 metres**

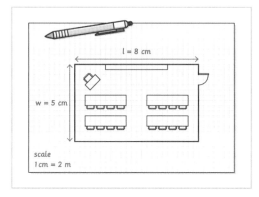

5. Emily draws a car with a scale of 1:20.
a b c What is the length of the car in metres?
Give your answer as a decimal and include the units m (metres).

- ▪ **4.5 metres** ▪ **4.5 m**

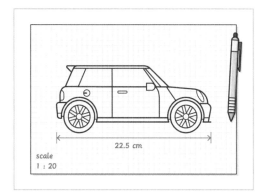

6. A map has a scale of 1 cm to 500 m.
   a b c What is the actual distance in kilometres from Skipwick to Malton?
   Give your answer as a decimal and include the units km (kilometres).

   ▪ 3.25 kilometres ▪ 3.25 km

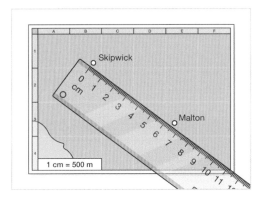

7. Halima draws a building with a scale of 1:250.
   a b c What is the actual width of the building in metres?
   Include the units m (metres) in your answer.

   ▪ 60 m ▪ 60 metres

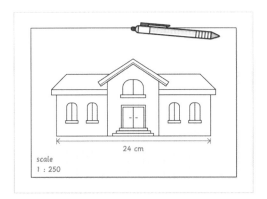

8. A design for a chair is drawn with a scale of 1 cm to 4 cm.
   a b c What is the actual height of the chair?
   Include the units cm (centimetres) in your answer.

   ▪ 112 cm ▪ 112 centimetres

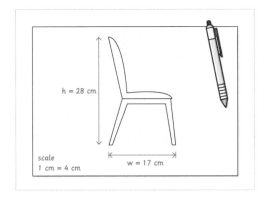

9. Jake draws a plan of his garden with a scale of 1:50.
   a b c What is the length of his garden in metres?
   Include the unit in your answer.

   ▪ 12 m ▪ 12 metres

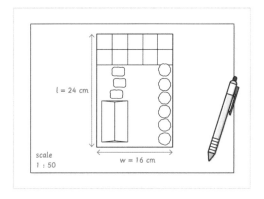

10. Alice draws a plan of her bedroom with a scale of 1 cm to 25 cm.
    a b c What is the width of her bedroom in metres?
    Include the units m (metres) in your answer.

    ▪ 4 m ▪ 4 metres

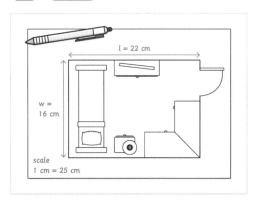

Level 2: Fluency - Finding scales and missing lengths on scale drawings.

❋ **Required:** 7/10   ❋ **Student Navigation:** on
❋ **Randomised:** off

11. A scale drawing is made of a room which is 5 metres long by 3 metres wide. How many centimetres does 1 centimetre on the drawing represent?
    a b c *Include the units cm (centimetres) in your answer.*

    ▪ 20 cm ▪ 20 centimetres

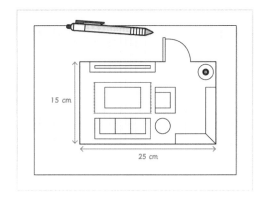

12. Duncan draws a 28 metre by 50 metre swimming
a pool. What scale has Duncan used?
b *Give your answer as a simplified ratio in the form*
c *1:n where 1 represents the distance on the*
drawing.

▪ 1:200

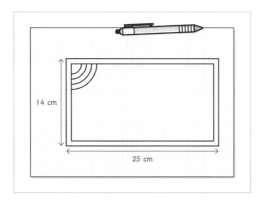

13. Corey draws a 1.8 metre long shed using a scale of
a 1 centimetre represents 5 centimetres. What is
b the missing length on the drawing?
c *Include the unit centimetres (cm) in your answer.*

▪ 36 cm ▪ 36 centimetres

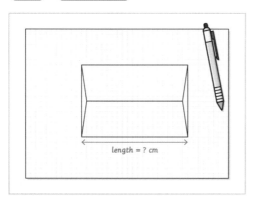

14. Erin draws a 1.2 metre wide cupboard with a scale
a of 1:4. What is the missing width on her drawing?
b *Include the units cm (centimetres) in your answer.*
c

▪ 30 cm ▪ 30 centimetres

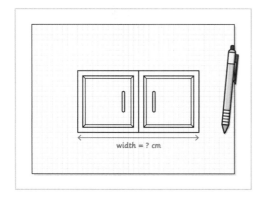

15. Tori draws a plan of a 15 metre by 12 metre
a classroom. What scale has Tori used?
b *Give your answer as a simplified ratio in the form*
c *1:n where 1 represents the distance on the*
drawing.

▪ 1:50

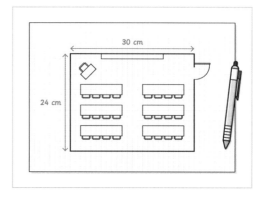

16. A 294 metre boat is drawn at a scale of 1
a centimetres represents 5 metres.
b How many centimetres long should the drawing
c be?
Give your answer as a decimal and include the
units cm (centimetres).

▪ 58.8 cm ▪ 58.8 centimetres

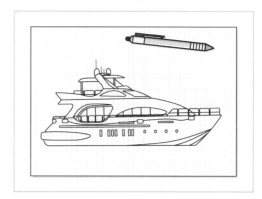

17. A map of an area with length 30 kilometres and
a width 20 kilometres is drawn at a scale of
b 1:50,000. What is the missing length on the map?
c *Include the units cm (centimetres) in your answer.*

▪ 60 cm ▪ 60 centimetres

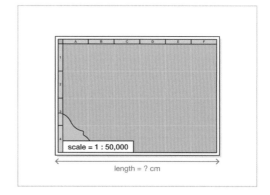

18. A 15 metre square play area is designed at a scale of 1:200. How many centimetres should each side of the square be on the drawing?
Include the units cm (centimetres) in your answer.

- 7.5 cm - 7.5 centimetres

19. Frankie draws a 180 centimetre long table with a scale of 1:5. How many centimetres long should his drawing be?
Include the units cm (centimetres) in your answer.

- 36 centimetres - 36 cm

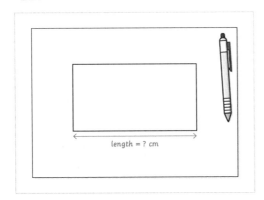

20. A church which is 170 metres high is drawn at a scale of 1:200. How many centimetres high should the drawing be?
Include the units cm (centimetres) in your answer.

- 85 centimetres - 85 cm

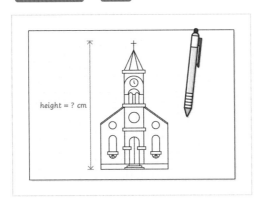

21. A drawing is made of a garden using a scale of 1:50.
What is the ratio of the **area** of the drawing to the area of the actual garden in the form 1:*n*?

- 1:2,500 - 1:2500

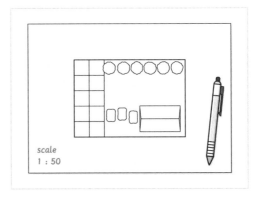

22. Jack is making a scale drawing of a football pitch which is 120 metres by 75 metres. Jack is drawing it on an A4 piece of paper which is 297 mm by 210 mm.
What would be a sensible scale for Jack to use? Explain your answer.

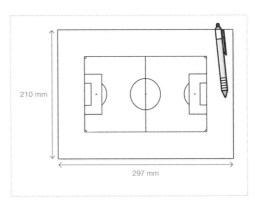

23. What is a reasonable estimate for the height of the tree?

- 1 - 5 metres - 5 - 10 metres - 10 - 15 metres

1/3

Level 3: cont.

24. Three maps are drawn using three different scales. If the maps cover the same region, sort the size of the maps in ascending order, smallest first.

- 1 cm on the map represents ¼ km. • 1:24,000
- 100 metres is represented by 5 mm on the map.

25. A plan of a desk is drawn to a scale of 1:5. If the desk is 180 cm longer than the drawing, how long is the drawing?
Include the units cm (centimetres) in your answer.

- 45 centimetres • 45 cm

Level 4: Problem Solving - Multi-step problems using scales.

✹ **Required:** 5/5   ✹ **Student Navigation:** on
✹ **Randomised:** off

26. A rectangular room is drawn using a scale of 1:50. The drawing of the room is 15 centimetres long and 12 centimetres wide. Carpet for the room costs £9.50 per square metre.
What is the cost of carpeting the room?
Include the £ sign in your answer.

- £427.50

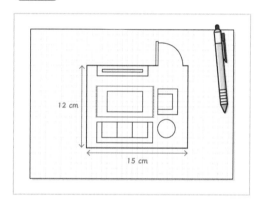

27. Yasmin is travelling from Caldwell to Stanmouth. If Yasmin drives at an average speed of 80 km/h, how many hours will the journey take her?
Give your answer as a decimal and don't include units.

- 2.5

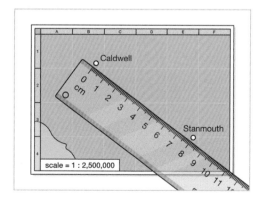

28. Here is a plan of the ground floor of a house. What is the area of the lounge in metres squared?
Don't include the unit in your answer.

- 56

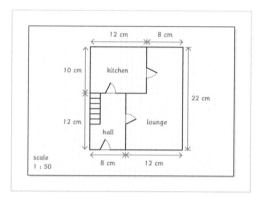

29. Tom wants to tile a wall in his kitchen. Each tile is 20 cm by 10 cm and they cost £5.95 for a pack of ten. How much will it cost to tile the wall?
Include the £ sign in your answer.

- £119 • £119.00

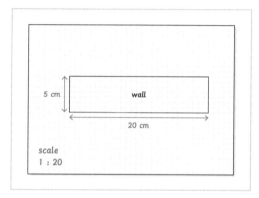

30. Sophie and Rabia are making a scale drawing of their school hall.
Sophie uses a scale of 1:40 and Rabia uses a scale of 1:50. Sophie's drawing is 10 centimetres longer than Rabia's. How long is the actual hall in metres?
Include the units m (metres) in your answer.

- 20 metres • 20 m

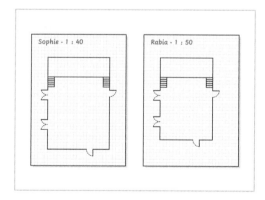

Write a Ratio in the Form 1:n and n:1

Competency: Use ratio notation, including reduction to simplest form.

Quick Search Ref: 10338

Correct: Correct.   **Wrong:** Incorrect, try again.   **Open:** Thank you.

Level 1: Understanding - Finding missing numbers in unit ratios.

Required: 7/10   **Student Navigation:** on   **Randomised:** off

---

1. What is a unit ratio?

- A ratio with just one part.
- A ratio where the difference between the parts is one.
1/3 ■ A ratio where one of the parts is equal to one.

$$1 : n$$

4. What missing number completes the ratio?

4:9 = 1:__
Give your answer as a mixed number fraction.

■ 2 1/4

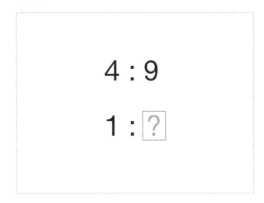
$$4 : 9$$
$$1 : \boxed{?}$$

2. What missing number completes the ratio?
2:3 = 1:__
Give your answer as a decimal.

■ 1.5

$$2 : 3$$
$$1 : \boxed{?}$$

5. What missing number completes the ratio?
7:4 = __:1
Give your answer as a decimal.

■ 1.75

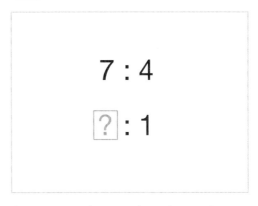
$$7 : 4$$
$$\boxed{?} : 1$$

3. What missing number completes the ratio?
5:4 = 1:__
Give your answer as a decimal.

■ 0.8

$$5 : 4$$
$$1 : \boxed{?}$$

6. What missing value completes the ratio?
2:9 = __:1
Give your answer as a fraction.

■ 2/9

$$2 : 9$$
$$\boxed{?} : 1$$

Level 1: *cont.*

7. What missing number completes the ratio?
    a b c 5:3 = __:1
    Give your answer a mixed number fraction.
    ▪ 1 2/3

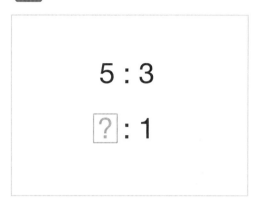

8. What missing number completes the ratio?
    1 2 3 2:11 = 1:__
    Give your answer as a decimal.
    ▪ 5.5

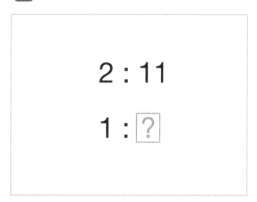

9. What missing number completes the ratio?
    1 2 3 5:2 = __:1
    Give your answer as a decimal.
    ▪ 2.5

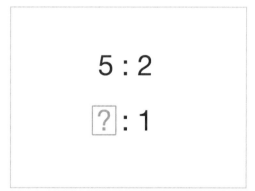

10. What missing number completes the ratio?
    a b c 5:7 = 1:__
    Give your answer as a mixed number fraction.
    ▪ 1 2/5

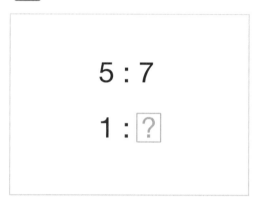

Level 2:   Fluency - Converting ratios to unit ratios.
    ✹ Required: 7/10   ✹ Student Navigation: on
    ✹ Randomised: off

11. Which **two** ratios are equivalent to 2:5?
    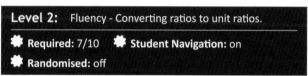
    ▪ 2.5:1 ▪ 1:2.5 ▪ 1:4 ▪ 0.4:1 ▪ 0.2:1
    2/5

12. What is the ratio 2:7 in the form 1:*n*?
    a b c *Give your answer as a decimal and use a colon (:)*
    to separate the values.
    ▪ 1:3.5

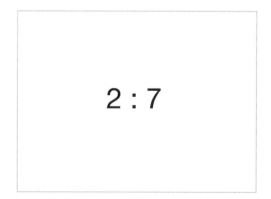

Level 2: *cont.*

13. What is the ratio 4:5 in the form of 1:*n*?

Give your answer as a decimal and use a colon (:) to separate the values.

▪ 1:1.25

$$4:5$$

14. What is the ratio 3:2 in the form *n*:1?

Give your answer as a mixed number fraction and use a colon (:) to separate the values.

▪ 1 1/2:1

$$3:2$$

15. What is the ratio ¼:3 in the form 1:*n*?
Use a colon (:) to separate the values.

▪ 1:12

$$\frac{1}{4}:3$$

16. What is 5:0.4 in the form *n*:1?

Give your answer as a decimal and use a colon (:) to separate the values.

▪ 12.5:1

$$5:0.4$$

17. In 9 carat gold the ratio of gold to other metals is 3:5. How many grams of other metals are there for every gram of gold?
Give your answer as a mixed number fraction. Don't include the unit.

▪ 1 2/3

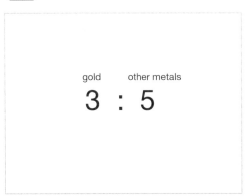
gold   other metals
$$3:5$$

18. What is the ratio 6:7 in the form *n*:1?

Give your answer as a fraction and use a colon (:) to separate the values.

▪ 6/7:1

$$6:7$$

19. What is the ratio 5:9 in the form 1:*n*?

a
b
c

Give your answer as a decimal and use a colon (:) to separate the values.

▪ 1:1.8

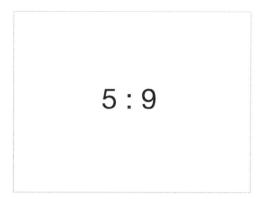

$$5 : 9$$

20. What is the ratio 4:13 in the form 1:*n*?

a
b
c

Give your answer as a mixed number fraction and use a colon (:) to separate the values.

▪ 1:3 1/4

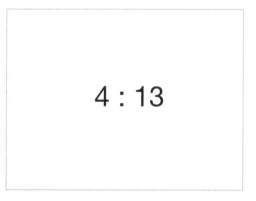

$$4 : 13$$

Level 3: Reasoning - Comparing amounts using unit ratios.

✿ **Required:** 5/5   ✿ **Student Navigation:** on
✿ **Randomised:** off

21. Robyn and Sapna are both making a drink. Robyn

a
b
c

mixes 40 ml (millilitres) of cordial with 500 ml of water.
Sapna mixes 25 ml of cordial with 300 ml of water. Whose drink tastes stronger? Explain your answer.

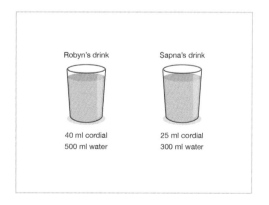

Robyn's drink   Sapna's drink

40 ml cordial   25 ml cordial
500 ml water   300 ml water

22. An orange paint is made by mixing 4 litres of red

a
b
c

paint with 5 litres of yellow paint.
How much yellow paint is needed to mix with 10 litres of red paint?
Give your answer as a decimal and include the units litres (l) in your answer.

▪ 12.5 litres ▪ 12.5 l

23. Three marine fish tanks contain the given mixtures

↑
↓

of salt and water.
Arrange the mixtures from most salty to least salty.

▪ mixture C ▪ mixture A ▪ mixture B

	salt in grams	water in litres
mixture A	67	20
mixture B	165	50
mixture C	136	40

24. A school trip must have a ratio of teachers to

a
b
c

students of at least 1:8.
Sumtown school plan to send 12 teachers and 102 students on a trip. Do they have enough teachers? Explain your answer.

25. Four shades of green paint are made by mixing

↑
↓

blue and yellow paint. The ratios of blue to yellow paint are given below.
Arrange the mixtures from lightest to darkest.

▪ 3:10 ▪ 4:13 ▪ 5:16 ▪ 1:3

Level 4: Problem Solving - Multi-step problems using unit ratios.

✿ **Required:** 5/5   ✿ **Student Navigation:** on
✿ **Randomised:** off

26. The grid shows nine ratios. There are four pairs of

a
b
c

equivalent ratios. Which is the odd one out in the form 1:n?
Give your answer as a decimal.

▪ 1:2.5

4 : 9	1 : 2.4	7 : 15
$1 : 2\frac{1}{7}$	5 : 12	$1 : 2\frac{1}{3}$
3 : 7	1 : 2.25	2 : 5

Level 4: *cont.*

27. 96 centimetres on a map measures 24 kilometres
a on land. What is the scale of the map in the form
b 1:*n* where 1 represents the distance on the map?
c

▪ `1:25,000` ▪ `1:25000`

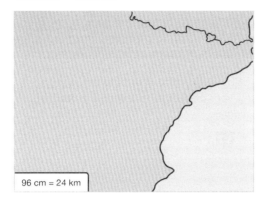

96 cm = 24 km

28. A van is 5.4 metres long and a scale model of it is
a 22.5 centimetres long. If the van is 1.92 metres
b wide, how wide is the model?
c *Include the units centimetres (cm) in your answer.*

▪ `8 centimetres` ▪ `8 cm`

29. If 5 miles are approximately equal to 8 kilometres,
a how many kilometres are there in 14 miles?
b *Give your answer as a decimal and include the*
c *units.*

▪ `22.4 kilometres` ▪ `22.4 km`

5 miles = 8 kilometres

14 miles = [?]

30. Fazel goes to Europe and exchanges £220 for
a €250. He returns with €30 and exchanges this for
b pounds at the same rate. How many pounds does
c Fazel receive?
Include the £ sign in your answer.

▪ `£26.40`

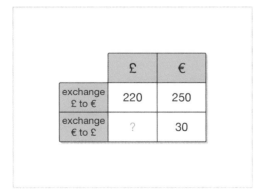

	£	€
exchange £ to €	220	250
exchange € to £	?	30

Write and Simplify Ratios

Competency: Use ratio notation, including reduction to simplest form.

Quick Search Ref: 10310

Correct: Correct.   Wrong: Incorrect, try again.   Open: Thank you.

Level 1: Understanding - Write and simplify basic ratios.

🌼 **Required:** 7/10   🌼 **Student Navigation:** on   🌼 **Randomised:** off

1. What is the ratio of apples to oranges?

abc *In your answer use a colon (:) to separate the values.*

▪ 2:5

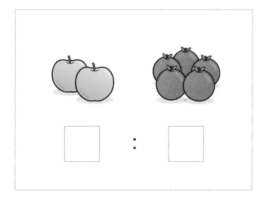

2. What is the ratio of marbles to buttons?

abc *Give your answer in its simplest form.*
Use a colon (:) to separate the values.

▪ 2:3

3. What is the ratio of footballs to tennis balls?

abc *Give your answer in its simplest form.*
Use a colon (:) to separate the values.

▪ 4:5

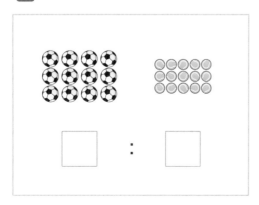

4. What is the ratio of pens to pencils?

abc *In your answer use a colon (:) to separate the values.*

▪ 5:2

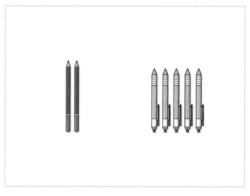

5. 7 out of 16 students are boys. What is the ratio of boys to girls?

abc *In your answer use a colon (:) to separate the values.*

▪ 7:9

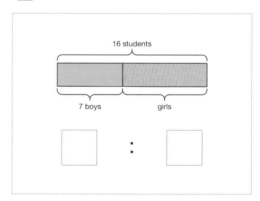

6. What is the ratio 5:100 in its simplest form?

1/4

▪ 0.5:10 ▪ ½:10 ▪ 1:20 ▪ 0.05:1

Level 1: *cont.*

7. A bus is carrying 48 passengers. 28 of the passengers are children. What is the ratio of children to adult passengers?
Give your answer in its simplest form.
Use a colon (:) to separate the values.

a
b
c

- 7:5

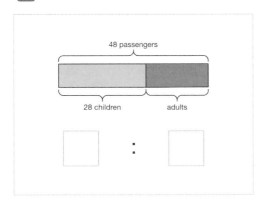

8. Year 8 has 48 girls and 54 boys. What is the ratio of girls to boys?
Give your answer in its simplest form.
Use a colon (:) to separate the values.

a
b
c

- 8:9

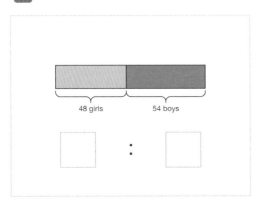

9. A bookshelf contains 120 books. 64 of the books are fiction. What is the ratio of fiction to non-fiction books?
Give your answer in its simplest form.
Use a colon (:) to separate the values.

a
b
c

- 8:7

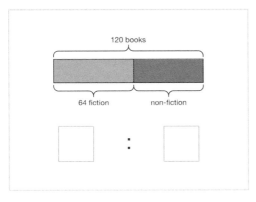

10. There are 40 cars and 28 bicycles in a school car park. What is the ratio of cars to bicycles in the car park?
Give your answer in its simplest form.
Use a colon (:) to separate the values.

a
b
c

- 10:7

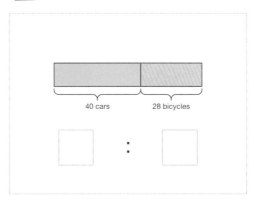

Level 2: Fluency - Three-part ratios and converting units.

✦ **Required:** 7/10   ✦ **Student Navigation:** on
✦ **Randomised:** off

11. A jug contains 150 millilitres of cordial and two litres of water. What is the ratio of cordial to water in the jug?
Give your answer in its simplest form.

a
b
c

- 3:40

12. Simplify the ratio 16:24:80.

a
b
c

- 2:3:10

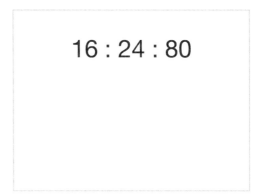

16 : 24 : 80

Level 2: cont.

13. A kitten has a mass of ½ a kilogram and its mother
a has a mass of 3 kilograms. What is the ratio of the
b kitten's mass to the mass of its mother?
c *Give your answer in its simplest form.*

▪ 1:6

kitten = $\frac{1}{2}$ kg   cat = 3 kg

14. One-third of a class are boys. What is the ratio of
a boys to girls in the class?
b *Give your answer in its simplest form.*
c

▪ 1:2

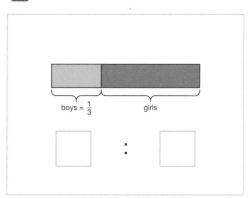

boys = $\frac{1}{3}$ girls

☐ **:** ☐

15. What is 0.4:3 as a whole number ratio in its
a simplest form?
b
c ▪ 2:15

0.4 : 3

16. Which ratio is equivalent to 3:2?

▪ 4:6 ▪ 31:21 ▪ 4:3 ▪ 18:12 ▪ 6:9

1/5

17. A recipe for shortcrust pastry contains 1 kilogram
a of plain flour, ½ kilogram of butter and 320 grams
b of water. What is the ratio of plain flour to butter
c to water?
Give your answer in its simplest form.

▪ 50:25:16

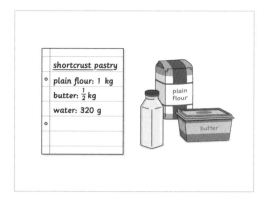

shortcrust pastry
○ plain flour: 1 kg
butter: $\frac{1}{2}$ kg
water: 320 g
○

18. A bag contains red and blue counters. 2/7 of the
a counters are red.
b What is the ratio of red counters to blue counters?
c *Give your answer in its simplest form.*

▪ 2:5

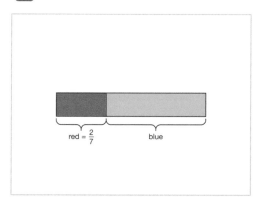

red = $\frac{2}{7}$ blue

19. Which ratio is equivalent to 3:8?

▪ 16:6 ▪ 2:7 ▪ 32:82 ▪ 24:64 ▪ 80:30

1/5

20. A mattress has a length of 2 metres, a width of 1.5
a metres and a height of 20 centimetres. What is the
b ratio of its length to width to height?
c *Give your answer in its simplest form.*

▪ 20:15:2

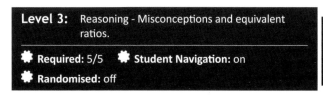

21. Sophie and Sadie are both making a drink.
a b c Sophie mixes 30 ml of cordial with 280 ml of water.
Sadie mixes 20 ml of cordial with 190 ml of water. Who has the stronger tasting cordial? Explain your answer.

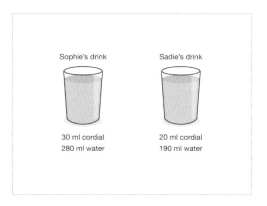

Sophie's drink   Sadie's drink

30 ml cordial   20 ml cordial
280 ml water   190 ml water

22. Shane has made a mistake in his simplifying ratios
a b c homework. What is the correct answer to the question he got wrong?

▪ 22:20:1

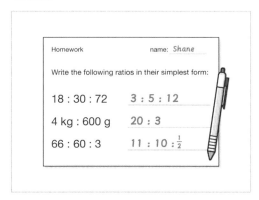

Homework   name: Shane

Write the following ratios in their simplest form:

18 : 30 : 72   3 : 5 : 12

4 kg : 600 g   20 : 3

66 : 60 : 3   11 : 10 : ½

23. The ratio of boys to girls is 3:4 in Year 7 and 3:5 in
a b c Year 8. Ffion says there must be more girls in Year 8. Is Ffion correct? Explain your answer.

24. Three shades of green paint are made by mixing
↑↓ blue paint and yellow paint. The ratio of blue to yellow paint is given below.
Arrange the mixtures from lightest to darkest.

▪ 3:7 ▪ 1:2 ▪ 2:3

25. One quarter of the children in a football team are
a b c girls. Selena says the ratio of girls to boys in the team is 1:4. Is Selena correct? Explain your answer.

26. Two angles meet at a point on a line. The angles
1 2 3 are in the ratio 1:4. Calculate the size of the largest angle.
Don't include units in your answer.

▪ 144

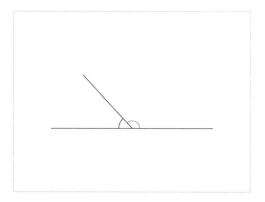

27. The grid shows nine ratios. There are four pairs of
a b c equivalent ratios. Which is the odd one out?
Give your answer in its simplest form.

▪ 2:5

6 : 8	2 : 3	9 : 15
30 : 35	8 : 20	3 : 4
10 : 15	6 : 10	12 : 14

28. Pink paint is made by adding 5 litres of white paint
1 2 3 to 8 litres of red paint.
Kian adds another 20 litres of red paint to the mix. How many litres of white paint does he need to add to keep the paint the same shade?
Give your answer as a decimal.
Don't include units in your answer.

▪ 12.5

Level 4: *cont.*

29. A bag contains blue, red and yellow counters.

a
b The ratio of blue counters to red is 3:2.
c The ratio of red counters to yellow is 3:4.
What is the ratio of blue counters to yellow?

▪ 9:8

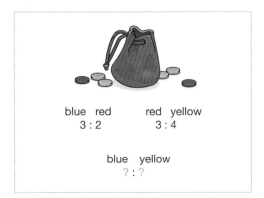

30. Two numbers are in the ratio 1:3. The difference
between the numbers is 8.
What is the product of the two numbers?

▪ 48

Convert Between Ratio and Proportion

Competency: Use ratio notation, including reduction to simplest form

Quick Search Ref: 10328

Correct: Correct.   **Wrong:** Incorrect, try again.   **Open:** Thank you.

Level 1: Understanding - Convert between ratio and proportion with numbers shown.

🌼 **Required:** 7/10   🌼 **Student Navigation:** on   🌼 **Randomised:** off

---

1. A ratio shows:

   1/4
   - the difference between one value and another value.
   - the size of one value in comparison to the whole.
   - the sum of one value and another value.
   - **the size of one value in comparison to another value.**

2. A proportion shows:
   1/4
   - the difference between one value and another value.
   - **the size of one value in comparison to the whole.**
   - the sum of one value and another value.
   - the size of one value in comparison to another value.

3. What is the ratio of cats to dogs?
   Give your answer in its simplest form.
   Use a colon (:) to separate the values.

   - 1:2

   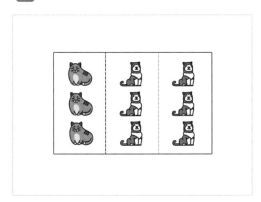

4. What proportion of the counters are white?
   Select two answers.

   2/6

   - 40% ▪ **1/5** ▪ 25% ▪ 1/4 ▪ **20%** ▪ 3/12

   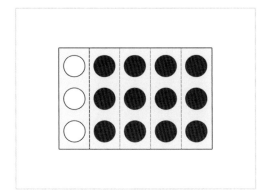

5. What proportion of the students are boys?
   Give your answer as a fraction in its simplest form.

   - 1/4

   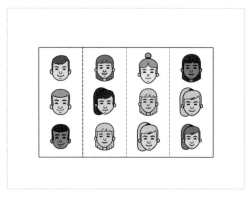

6. One-third of the balls are footballs. What is the ratio of footballs to tennis balls?
   Give your answer in its simplest form.
   Use a colon (:) to separate the values.

   - 1:2

   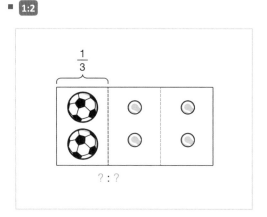

7. Two-fifths of the birds are ducks. What proportion of the birds are ducks?
   Give your answer as a percentage and include the % sign.

   - 40%

   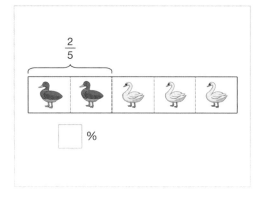

8. One-fifth of the shapes are squares. What is the ratio of squares to circles?

a
b
c

Give your answer in its simplest form.
Use a colon (:) to separate the values.

▪ 1:4

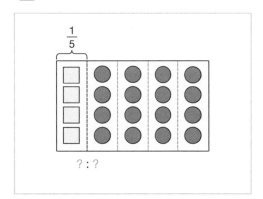

9. The ratio of apples to oranges is 1:5. What proportion of the fruit are apples?

a
b
c

Give your answer as a fraction in its simplest form.

▪ 1/6

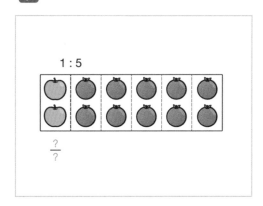

10. The ratio of large beads to small beads is 2:3. What proportion of the beads are large?

a
b
c

Give your answer as a percentage and include the % sign.

▪ 40%

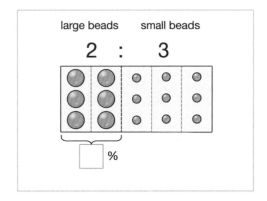

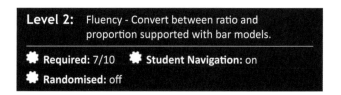

Level 2:   Fluency - Convert between ratio and proportion supported with bar models.

✹ Required: 7/10   ✹ Student Navigation: on
✹ Randomised: off

11. The ratio of men to women at a concert is 1:5. What proportion of the audience are male?

a
b
c

Give your answer as a fraction.

▪ 1/6

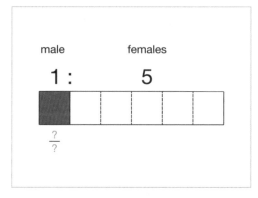

12. A bag contains green and yellow marbles. One-eighth of the marbles are green. What is the ratio of green to yellow marbles?

a
b
c

Use a colon (:) to separate the values.

▪ 1:7

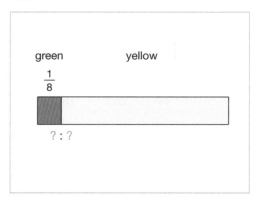

13. A farmer has cows and sheep in the ratio 2:3. What proportion of the animals are cows?

a
b
c

Give your answer as a percentage and include the % sign.

▪ 40%

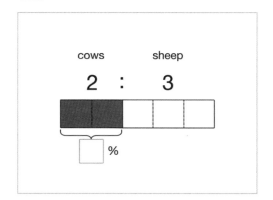

14. A box contains black and red pens. Four-ninths of the pens are black. What is the ratio of black pens to red pens?
Use a colon (:) to separate the values.

a
b
c

▪ **4:5**

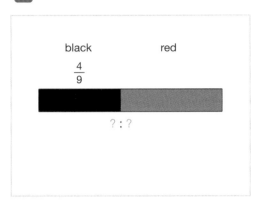

15. Emily makes a drink with 2 parts cordial to every 11 parts water. What proportion of the drink is water?
Give your answer as a fraction.

a
b
c

▪ **11/13**

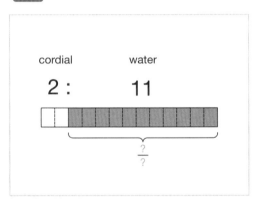

16. 3/7 of students in a class are boys. What is the ratio of girls to boys?
Use a colon (:) to separate the values.

a
b
c

▪ **4:3**

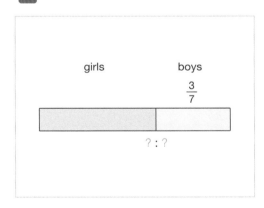

17. A bag contains blue and red counters in the ratio 3:5. What proportion of the counters are blue?
Give your answer as a fraction.

a
b
c

▪ **3/8**

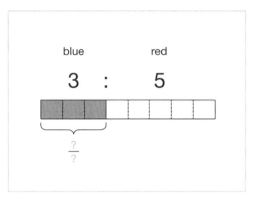

18. A purse contains silver and copper coins. Two-sevenths of the coins are silver. What is the ratio of silver to copper coins?
Use a colon (:) to separate the values.

a
b
c

▪ **2:5**

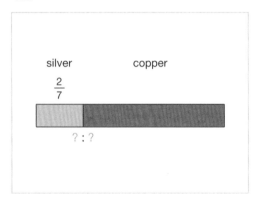

19. Joseph has black and white socks in the ratio 3:1. What proportion of his socks are black?
Give your answer as a percentage and include the % sign.

a
b
c

▪ **75%**

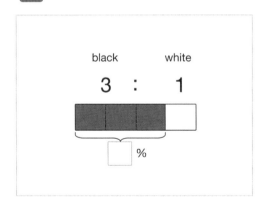

Level 2: *cont.*

20. A sports coach has white and yellow cones. Seven-tenths of her cones are yellow. What is the ratio of white to yellow cones?
Use a colon (:) to separate the values.

a
b
c

■ **3:7**

Level 3: Reasoning - Misconceptions and reasoning with ratio and proportion.

✱ **Required:** 5/5   ✱ **Student Navigation:** on
✱ **Randomised:** off

21. Archie, Bradley and Charlie split their stickers in the ratio 2:3:4.
What fraction of the stickers does Charlie receive?

a
b
c

■ **4/9**

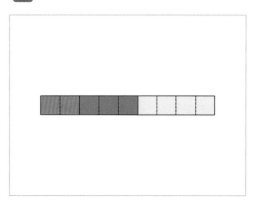

22. The ratio of boys to girls in Year 7 is 3:4. Sophie says that one-quarter of Year 7 must be girls. Is Sophie right? Explain your answer.

a
b
c

23. A light blue paint is made by mixing white and blue paint. Which mixture gives the lightest colour paint?

1/4   ■ mixture A ■ mixture B ■ mixture C ■ mixture D

A white : blue 1 : 3	B $\frac{1}{3}$ white
C 70% blue	D blue : white 5 : 2

24. A junior athletics club has twice as many girls as boys. What fraction of the members are boys?

a
b
c

■ **1/3**

25. A bag contains green and yellow counters. Three of the following statements are true, two are false. Select the **two false** statements.

2/5   ■ 40% of the counters are yellow
■ The ratio of green counters to yellow is 2:3.
■ Three-fifths of the counters are yellow.
■ Two-thirds of the counters are green
■ There are 24 yellow counters and 16 green counters.

Level 4: Problem Solving - Multi-step problems and changes in ratio and proportion.

✱ **Required:** 5/5   ✱ **Student Navigation:** on
✱ **Randomised:** off

26. A bag contains blue, red and yellow counters. One-quarter of the counters are blue and the ratio of red to yellow counters is 2:1. What is the ratio of blue to red to yellow counters?

a
b
c

■ **1:2:1**

27. At the cinema, one-quarter of the audience are men, one-third are women and the rest are children. What is the ratio of men to women to children?

a
b
c

■ **3:4:5**

28. The ratio of males to females on a bus is 2:3. At the next stop 6 men get off the bus. Two-thirds of the remaining passengers are female. How many people are now on the bus?

1
2
3

■ **54**

29. The grid shows nine bags containing red and blue
counters. Each bag shows either the ratio of blue
to red counters or the proportion of blue counters
inside. Four pairs of bags have the same
proportion of blue counters. What is the
proportion of blue counters in the odd one out?
Give your answer as a fraction.

a
b
c

■ `3/8`

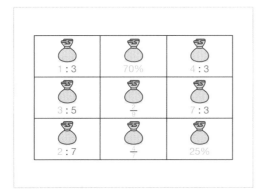

30. Lulu is one-third of the age of Katie. In ten years
time, their ages will be in the ratio 3:4. How many
years old is Katie now?

1
2
3

■ `6`

Direct Proportion Using "Best Buys"

Competency: Solve problems involving direct and inverse proportion, including graphical and algebraic representations.

Quick Search Ref: 10415

Correct: Correct.   Wrong: Incorrect, try again.   Open: Thank you.

Level 1: Understanding - Using proportion to find the cost of items.

✸ **Required: 7/10**   ✸ **Student Navigation: on**   ✸ **Randomised: off**

1. 1 pencil costs 12 pence. What is the cost of 5 pencils?
a
b *Include the units p (pence) in your answer.*
c

■ 60p ■ 60 pence

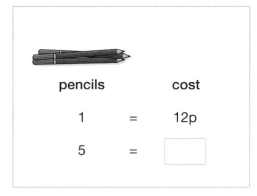

pencils		cost
1	=	12p
5	=	

2. If 7 t-shirts cost £56, how much is 1 t-shirt?
a *Include the £ (pound) sign in your answer.*
b
c ■ £8.00 ■ £8

t-shirts		cost
7	=	£56
1	=	

3. 4 yoghurts cost £1.30. How much are 12 yoghurts?
a *Include the £ (pound) sign in your answer.*
b
c ■ £3.90

yogurts		cost
4	=	£1.30
12	=	

4. If 6 loaves of bread cost £7.20, what do 2 loaves cost?
a
b *Include the £ (pound) sign in your answer.*
c

■ £2.40

loaves		cost
6	=	£7.20
2	=	

5. 10 eggs cost £2.80. What is the cost of 15 eggs?
a *Include the £ (pound) sign in your answer.*
b
c ■ £4.20

eggs		cost
10	=	£2.80
5	=	
15	=	

6. If 12 apples cost £3.36, how much are 8 apples?
a *Include the £ (pound) sign in your answer.*
b
c ■ £2.24

apples		cost
12	=	£3.36
4	=	
8	=	

Level 1: *cont.*

7. 16 cookies cost £2.24. How much do 20 cookies cost?
a b c *Include the £ (pound) sign in your answer.*

▪ £2.80

cookies		cost
16	=	£2.24
4	=	
20	=	

8. If 3 pens cost £1.32, what do 9 pens cost?
a b c *Include the £ (pound) sign in your answer.*
▪ £3.96

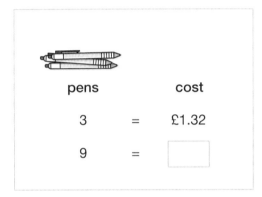

pens		cost
3	=	£1.32
9	=	

9. 12 stickers cost £1.50. What is the cost of 18 stickers?
a b c *Include the £ (pound) sign in your answer.*
▪ £2.25

stickers		cost
12	=	£1.50
6	=	
18	=	

10. If 16 buns cost £2.40, how much do 12 buns cost?
a b c *Include the £ (pound) sign in your answer.*
▪ £1.80

buns		cost
16	=	£2.40
4	=	
12	=	

Level 2: Fluency - Finding best buys including the unitary method.

✱ **Required:** 7/10   ✱ **Student Navigation:** on
✱ **Randomised:** off

11. A supermarket sells a single pepper for 55p or a pack of 3 peppers for 99p.
a b c What is the saving per pepper when buying a pack of 3 peppers?
Include the units p (pence) in your answer.

▪ 22p ▪ 22 pence

peppers

3 for 99p   1 for 55p

12. A theme park charges an entry fee of £29 per person or £100 for a group of 4.
a b c How much can a 4 people save in total when buying a group ticket?
Include the £ (pound) sign in your answer.

▪ £16.00 ▪ £16

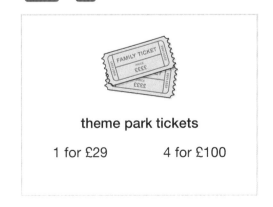

theme park tickets

1 for £29   4 for £100

13. A shop sells a pack of 2 notepads for £1.20 or a
a pack of 3 notepads £1.59.
b What is the saving per notepad when buying a
c pack of 3 notepads?
Include the units p (pence) in your answer.

▪ 7p ▪ 7 pence

notepads

2 for £1.20   3 for £1.59

1 for [   ]   1 for [   ]

14. A supermarket sells a pack of 2 fishcakes for £1.65
a or a pack of 6 for £3.90.
b What is the saving on 2 fishcakes when bought as
c part of a pack of 6?
Include the units p (pence) in your answer.

▪ 35 pence ▪ 35p

fishcakes

2 for £1.65   6 for £3.90

                     2 for [   ]

15. A sports shop sells tennis balls in a pack of 3 for
a £9.99 or a pack of 10 for £25.
b What is the saving per tennis ball when buying a
c pack of 10?
Include the units p (pence) in your answer.

▪ 83 pence ▪ 83p

tennis balls

3 for £9.99   10 for £25

1 for [   ]   1 for [   ]

16. Smart Buy sells packs of 2 pairs of socks for £3.15.
a Value Fashion charge £4.75 for 3 pairs of the same
b socks.
c How much more expensive is it to buy 6 pairs from
Value Fashion than Smart Buy?
Include the units p (pence) in your answer.

▪ 5 pence ▪ 5p

socks

Smart Buy   Value Fashion

2 for £3.15   3 for £4.75

6 for [   ]   6 for [   ]

17. Green Hands garden centre sell 8 plants for £7.70.
a Spades of Choice sell 12 plants for £9.90.
b What is the saving on 4 plants from Spades of
c Choice?
Include the units p (pence) in your answer.

▪ 55 pence ▪ 55p

plants

Green Hands   Spades of Choice

8 for £7.70   12 for £9.90

4 for [   ]   4 for [   ]

18. A hotel charges £350 for a 7 night stay or £170 for
a a 2 night stay.
b What is the saving per night when booking for a
c week?
Include the £ (pound) sign in your answer.

▪ £35 ▪ £35.00

nights stay at a hotel

7 for £350   2 for £170

1 for [   ]   1 for [   ]

Level 2: *cont.*

19. The Model Shop sells packs of 2 toy cars for £3.25 and The Toy Store sells packs of 3 cars for £4.75. How much cheaper is it to buy 6 cars from The Toy Store than The Model Shop?
Include the units p (pence) in your answer.

a
b
c

- 25 pence - 25p

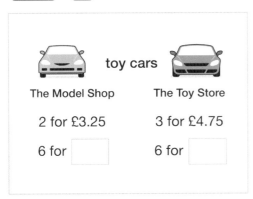

toy cars

The Model Shop   The Toy Store

2 for £3.25   3 for £4.75

6 for [ ]   6 for [ ]

20. Cans of cola are sold in packs of 18 for £6.99 or packs of 12 for £4.90. What is the saving on 6 cans when bought in packs of 18?
Include the units p (pence) in your answer.

a
b
c

- 12p - 12 pence

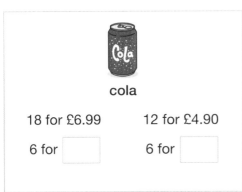

cola

18 for £6.99   12 for £4.90

6 for [ ]   6 for [ ]

Level 3: Reasoning - Selecting appropriate methods to compare prices.
_____

✿ **Required:** 5/5   ✿ **Student Navigation:** on
✿ **Randomised:** off

21. Which is the easiest method to find the best value for money for these packets of jam tarts?

a
b
c

jam tarts

8   for £1.22
12   for £2.10
20   for £4.15

22. Three footballers were bought at the start of the season by three different clubs. Based on cost per goal, sort the players according to value for money (best value first).

↑
↓

- Pablo Poacher - Wayne Worldie - Harry Hotshot

player	cost (£m)	goals
Harry Hotshot	8	30
Pablo Poacher	0.5	3
Wayne Worldie	3.5	14

23. Caleb is comparing prices of two taxi companies. Caleb says the cost per mile at Kwik Kabs is less so they always give the best value for money. Is Caleb correct? Explain your answer.

a
b
c

taxi prices

Kwik Kabs
£2.00 per trip plus 40p per mile

Speedy Cars
80p per mile

24. Arrange the following packs of toilet roll according to value for money (best value first).

↑
↓

- 9 for £4.19 - 10 for £4.99 - 4 for £2.29
- 2 for £1.20

25. Lois is comparing prices of bottles of milk. Lois says instead of calculating the cost per millilitre, she can calculate how many millilitres she can buy per penny. Is Lois correct? Explain your answer.

a
b
c

2,272 ml   for   £2
568 ml   for   71p

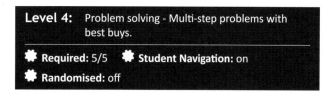

Level 4: Problem solving - Multi-step problems with best buys.

✸ **Required:** 5/5 ✸ **Student Navigation:** on
✸ **Randomised:** off

26. Milk is sold in two different sizes of bottle.
a A 4 pint bottle of milk costs £1.52.
b A 6 pint bottle of milk costs £2.16.
c How much does the 4 pint bottle have to be reduced by for it to be the same value for money?
Include the units p (pence) in your answer.

▪ 8p ▪ 8 pence

27. Cola is sold at five different shops. Arrange the prices in order of value for money (best value first).

▪ 1½ litre bottle for £1.89 ▪ 2 litre bottle for £2.55
▪ 12 × 330 ml cans for £6.00 ▪ 500 ml bottle for 99p
▪ 330 ml can for 70p

28. Ketchup is sold in two different sizes of bottle.
a A 354 g bottle of ketchup costs 90p.
b A 1,500 g bottle of ketchup costs £3.00.
c How much saving is made per 100 g when buying the large bottle?
Give your answer in pence rounded to the nearest penny.

▪ 5 pence ▪ 5p

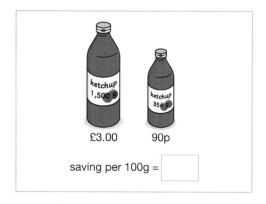

saving per 100g = ☐

29. A shop sells coffee in two different sizes of jar.
a A 150 g jar of coffee costs £4.59.
b A 275 g jar of coffee costs £8.25.
c What is the saving per 100 g when buying the large jar?
Include the units p (pence) in your answer.

▪ 6 pence ▪ 6p

saving per 100g = ☐

30. A supermarket sells two blocks of cheddar cheese.
a How many more grams per £1 do you receive with
b the larger block?
c Include the unit grams (g) in your answer.

▪ 3 g

Direct Proportion Using Currency

Competency: Solve problems involving direct and inverse proportion, including graphical and algebraic representations.

Quick Search Ref: 10414

Correct: Correct.   Wrong: Incorrect, try again.   Open: Thank you.

Level 1: Understanding - Currency conversions, non-calculator.

⚙ Required: 7/10   ⚙ Student Navigation: on   ⚙ Randomised: off

---

1. If the exchange rate between British pounds (£) and euros (€) is £1 = €1.10, what is the correct calculation to convert £56 to euros?

    1/4   ■ 1.10 + 56   ■ **56 × 1.10**   ■ 1.10 ÷ 56   ■ 56 ÷ 1.10

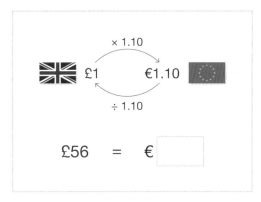

2. If the exchange rate between British pounds (£) and US dollars ($) is £1 = $1.40, how much is £200 worth in dollars?
Include the $ sign in your answer.

    ■ **$280.00**   ■ **$280**

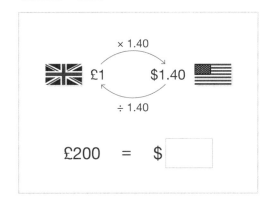

3. If the exchange rate between British pounds (£) and euros (€) is £1 = €1.20, how much is £150 worth in euros?
Don't include the € sign in your answer.

    ■ **180**

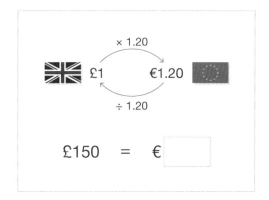

4. If the exchange rate between US dollars ($) and euros (€) is $1 = €0.80, how much is $40 worth in euros?
Don't include the € sign in your answer.

    ■ **32**

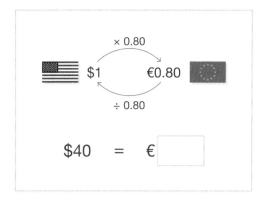

5. If the exchange rate between British pounds (£) and US dollars ($) is £1 = $1.40, how much is $420 worth in pounds?
Include the £ sign in your answer.

a b c

- £300.00 ▪ £300

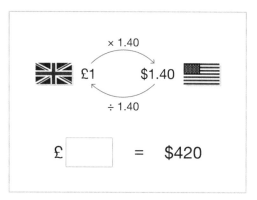

6. If the exchange rate between British pounds (£) and euros (€) is £1 = €1.10, how much is €550 worth in pounds?
Include the £ sign in your answer.

a b c

- £500 ▪ £500.00

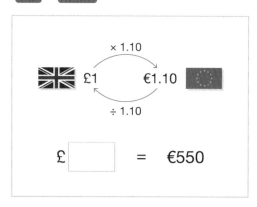

7. If the exchange rate between euros (€) and British pounds (£) is €1 = £0.80, how much is £160 worth in euros?
Don't include the € sign in your answer.

1 2 3

- 200

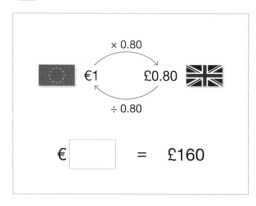

8. If the exchange rate between British pounds (£) and US dollars ($) is £1 = $1.40, how much is £70 worth in dollars?
Include the $ sign in your answer.

a b c

- $98.00 ▪ $98

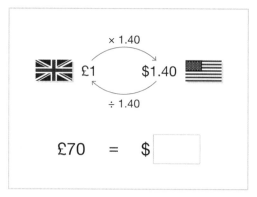

9. If the exchange rate between US dollars ($) and euros (€) is $1 = €0.80, how much is $360 worth in euros?
Don't include the € sign in your answer.

1 2 3

- 288

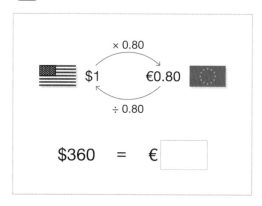

10. If the exchange rate between British pounds (£) and euros (€) is £1 = €1.20, how much is €30 worth in pounds?
Include the £ sign in your answer.

a b c

- £25.00 ▪ £25

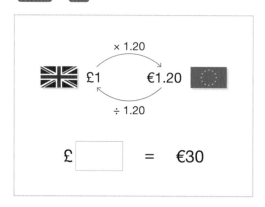

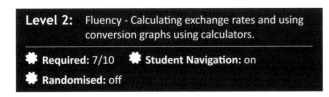

Level 2: Fluency - Calculating exchange rates and using conversion graphs using calculators.

⚙ **Required:** 7/10   ⚙ **Student Navigation:** on
⚙ **Randomised:** off

11. If the exchange rate between British pounds (£) and Japanese yen (¥) is £1 = ¥147, how much is £56 pounds worth in yen?
Don't include the unit in your answer.

a
b
c

▪ 8232 ▪ 8,232

£1 = ¥147

£56 = ¥ ☐

12. Olivia exchanges £250 for €275. How many euros did she receive for every £1?
Give your answer to 2 decimal places.
Don't include the € sign in your answer.

1
2
3

▪ 1.1

£250 = €275

£1 = € ☐

13. The graph shows the exchange rate between the British pound and the Indian rupee.
Use the graph to calculate the value of £500 in Indian rupee.
Don't include the units in your answer.

a
b
c

▪ 4,500 ▪ 4500

exchange rate between the British pound and the Indian rupee

Indian rupe (y-axis): 0, 1,000, 2,000, 3,000, 4,000, 5,000
British pound (£) (x-axis): 0, 100, 200, 300, 400, 500, 600

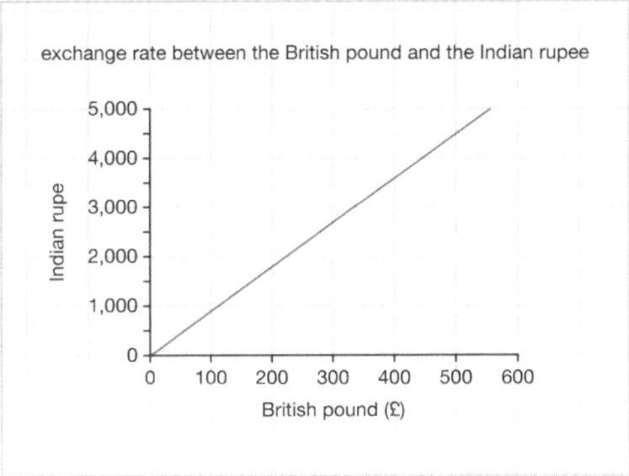

14. If the exchange rate between British pounds (£) and Australian dollars ($) is £1 = $1.77, how much is $200 worth in pounds?
Give your answer to 2 decimal places and include the £ sign.

a
b
c

▪ £112.99

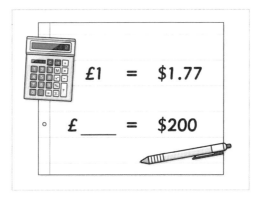

£1 = $1.77

£___ = $200

15. Kyra exchanges £150 for $250 Australian dollars. How many pounds did she receive for every dollar?
Give your answer to 2 decimal places and include the £ sign.

a
b
c

▪ £0.60

£150 = $250

£ ☐ = $1

16. The graph shows the exchange rate between the British pound and the US dollar.
Use the graph to calculate the value of $700 in pounds.
Include the £ sign in your answer.

a
b
c

▪ £500

exchange rate between the British pound and the US dollar

US dollar ($) (y-axis): 0, 100, 200, 300, 400, 500
British pounds (£) (x-axis): 0, 100, 200, 300, 400

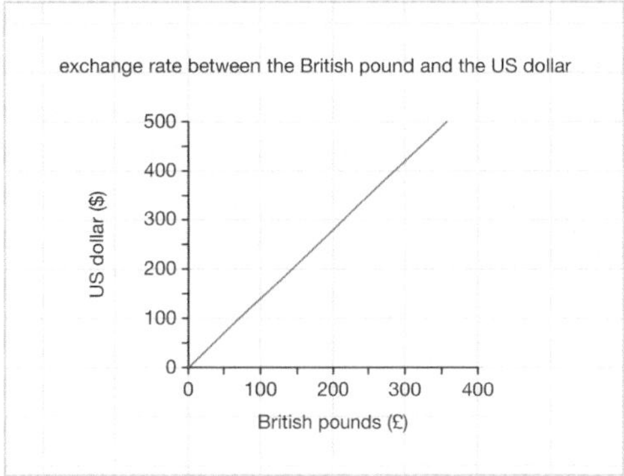

17. If the exchange rate between euros (€) and US
dollars ($) is €1 = $1.23, how much is €95 euros
worth in dollars?
*Give your answer to 2 decimal places and include
the $ sign.*

■ $116.85

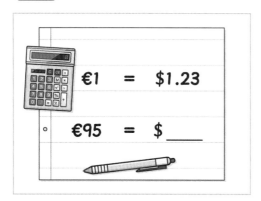

18. Aaqil exchanges £160 for $208 US dollars. How
many dollars did he receive for every £1?
*Give your answer to 2 decimal places and include
the $ sign.*

■ $1.30

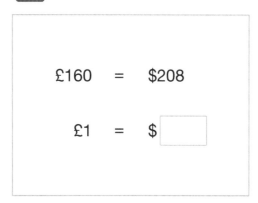

19. The graph shows the exchange rate between the
euro and the US dollar.
Use the graph to calculate the value of €500 in
dollars.
Include the $ sign in your answer.

■ $575

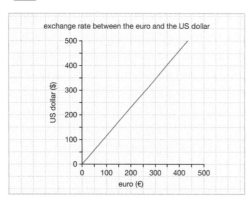

20. If the exchange rate between US dollars ($) and
British pounds (£) is $1 = £0.71, how much is £85
worth in dollars?
*Give your answer to 2 decimal places and include
the $ sign.*

■ $119.72

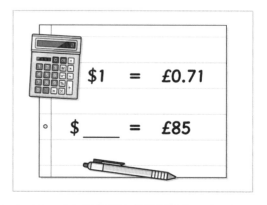

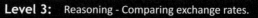

✹ **Required:** 5/5 ✹ **Student Navigation:** on
✹ **Randomised:** off

21. Mark can either exchange British pounds to US
dollars in England at a rate of £1 = $1.40 or in
America at a rate of $1 = £0.70. Describe the
method Mark could use to decide where he
should exchange his money.

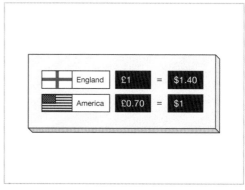

22. If the exchange rate between US dollars ($) and
euros is (€) is $1 = €0.80, how many dollars equal
1 euro?
*Give your answer as a decimal and include the $
sign.*

■ $1.25

Level 3: *cont.*

23. Helen is going on holiday and wants to exchange
a £420 to Turkish lira.
b The exchange rate is £1 = 5.40 Turkish lira.
c What is the most amount of 100 Turkish lira notes
Helen can get?

▪ 22

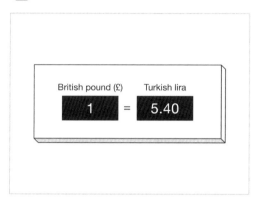

24. If £1 = €1.10 and €1 = $1.20, how many dollars
a does £1 equal?
b *Include the $ sign in your answer.*
c

▪ $1.32

25. Sort the exchange rates according to which rate
will give the most amount of euros per pound,
starting with most euros first.

▪ €1 = £0.85 ▪ €1 = €1.15 ▪ €1 = £0.90

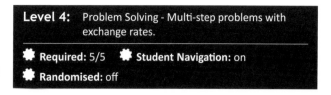

Level 4: Problem Solving - Multi-step problems with
exchange rates.
──────────────
✹ **Required:** 5/5 ✹ **Student Navigation:** on
✹ **Randomised:** off

26. Jason is travelling to France by plane and buys 2
1 sandwiches, a tea and a fruit juice. He pays part of
2 the cost with a £5 note and the remainder in
3 euros.
How much does Jason have to pay in euros?
Don't include the € sign in your answer.

▪ 7.26

27. Steph is in New York and sees some jeans for
a $49.95. She has seen the same jeans on sale in the
b UK for £40. If the exchange rate is £1 = $1.35, how
c much in British pounds could Steph save buying
her jeans in New York?
Include the £ sign in your answer.

▪ £3 ▪ £3.00

28. Talal sees a car in America for $35,000 that
a normally costs £28,000 in the UK. If the exchange
b rate is £1 = $1.40 and it costs £1,200 for Talal to
c import the car to the UK, how much does he save
buying it in America?
Give your answer in pounds and include the £ sign.

▪ £1800 ▪ £1800.00 ▪ £1,800.00 ▪ £1,800

29. Olivia is in Paris and sees some perfume for €46.95. She has seen the same perfume on sale in the UK for £44.40. If the exchange rate is £1 = €1.15, how much in euros could Olivia save buying her perfume in Paris?
Don't include the € sign in your answer.

▪ 4.11

UK   Paris

£1   =   €1.15

£44.40   €46.95

x
perfume
x

x
perfume
x

30. Chloe exchanges £350 to euros for a holiday at a rate of £1 = €1.20. She then decides to take a credit card instead so changes her money back to pounds at a different rate. Chloe receives £357 at the new rate. How much is one euro worth at the new rate?
Include the unit pence (p) in your answer.

▪ 85p ▪ 85 pence

Direct Proportion Using Recipes

Competency: Solve problems involving direct and inverse proportion, including graphical and algebraic representations.

Quick Search Ref: 10367

Correct: Correct.   Wrong: Incorrect, try again.   Open: Thank you.

Level 1: Recipe questions using proportional reasoning.

✦ Required: 10/10   ✦ Student Navigation: on   ✦ Randomised: off

---

1. Michael wants to make 30 cookies. How much butter does he need?
Include the unit g (grams) in your answer.

a b c

▪ 600 grams ▪ 600 g

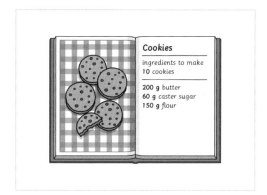

2. Danielle wants to make soup for 20 people. How many tomatoes does she need?

1 2 3

▪ 30

3. Darcy wants to make 12 pancakes. How much milk does she need?
Include the unit ml (millilitres) in your answer.

a b c

▪ 360 millilitres ▪ 360 ml

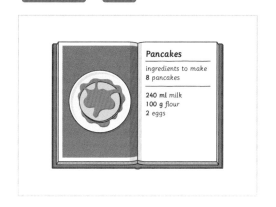

4. Carmen wants to make spaghetti for five people. How much plain flour does she need?
Include the unit g (grams) in your answer.

a b c

▪ 400 grams ▪ 400 g

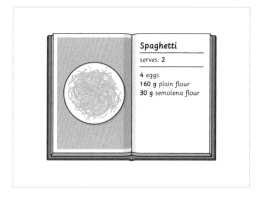

5. Aman wants to make 18 cupcakes. How much caster sugar does he need?
Include the unit g (grams) in your answer.

a b c

▪ 270 grams ▪ 270 g

6. Tilly wants to make 16 flapjacks. How much golden syrup does she need?
Include the unit ml (millilitres) in your answer.

a b c

▪ 48 ml ▪ 48 millilitres

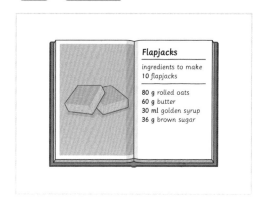

---

7. Lola wants to make macaroni cheese for seven people. How much cheese does she need?
a
b *Include the unit g (grams) in your answer.*
c

■ 350 grams ■ 350 g

8. Craig wants to make 20 cookies. How much caster sugar does he need?
a
b *Include the unit g (grams) in your answer.*
c

■ 120 g ■ 120 grams

9. Maria wants to make soup for ten people. How much vegetable stock does she need?
a
b *Include the unit l (litres) in your answer.*
c

■ 2.5 l ■ 2.5 litres

10. Fabienne wants to make ten pancakes. How much flour does she need?
a
b *Include the unit g (grams) in your answer.*
c

■ 125 g ■ 125 grams

Mathematics **Y8**

Algebra

Formulae

Expand and Factorise

Simplify

Solving Linear Equations

Inequalities

Rearrange Formulae to Change the Subject

Competency: Understand and use standard mathematical formulae: rearrange formulae to change the subject.

Quick Search Ref: 10248

Correct: Correct.   **Wrong:** Incorrect, try again.   **Open:** Thank you.

Level 1:   Understanding - Simple one or two step operations with two variables.

✿ **Required:** 7/10   ✿ **Student Navigation:** on   ✿ **Randomised:** off

---

1. Select the three equations that are equivalent to $a + 7 = b$.

▪ b + 7 = a ▪ b = 7 + a ▪ b - 7 = a ▪ a = 7 - b ▪ a = b - 7

3/5

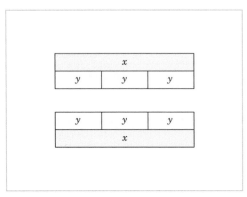

2. Select the three equations that are are equivalent to $3y = x$.

▪ x = 3y ▪ y = x/3 ▪ x = y/3 ▪ x/3 = y ▪ 3x = y

3/5

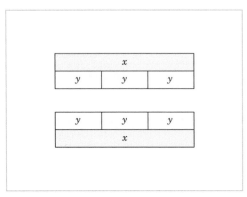

3. If $y = x/5$, which equation has been rearranged to make x the subject?

▪ x = 5y ▪ x = y/5 ▪ x/5 = y

1/3

$$y = \frac{x}{5}$$

$\downarrow \boxed{\times 5} \quad \downarrow \boxed{\times 5}$

$$5y = x$$

4. If $a = 2b + 3$, which equation has been rearranged to make b the subject?

▪ 2b = a - 3 ▪ b = a/2 - 3 ▪ b = (a - 3)/2

1/3

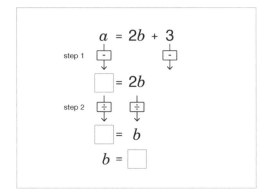

$$a = 2b + 3$$

step 1 $\boxed{-}$ $\boxed{-}$

$$\boxed{} = 2b$$

step 2 $\boxed{\div}$ $\boxed{\div}$

$$\boxed{} = b$$

$$b = \boxed{}$$

5. Rearrange the equation $a = b/4 - 2$ to make b the subject.

▪ b/4 = a + 2 ▪ b = 4(a + 2) ▪ b = 4a + 2

1/3

$$a = \frac{b}{4} - 2$$

step 1 $\boxed{+}$ $\boxed{+}$

$$\boxed{} = \frac{b}{4}$$

step 2 $\boxed{\times}$ $\boxed{\times}$

$$\boxed{} = b$$

$$b = \boxed{}$$

6. Make r the subject of the formula $s = 5/r$.

▪ r = 5/s ▪ 5/s = r

$$s = \frac{5}{r}$$

step 1 $\boxed{\times}$ $\boxed{\times}$

$$\boxed{} = 5$$

step 2 $\boxed{\div}$ $\boxed{\div}$

$$r = \boxed{}$$

Level 1: *cont.*

7. Rearrange the formula $s = (t - 5)/3$ to make t the
subject.

a
b
c   ▪ t = 5 + 3s ▪ 3s + 5 = t ▪ t = 3s + 5 ▪ 5 + 3s = t

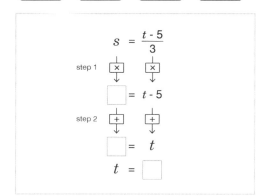

8. Make a the subject of $b = (a + 7)/3$.

a
b
c   ▪ -7 + 3b = a ▪ a = 3b - 7 ▪ a = -7 + 3b ▪ 3b - 7 = a

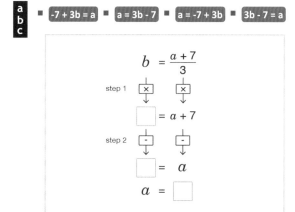

9. Rearrange $d = 5/e$ to make e the subject of the
formula.

a
b
c   ▪ e = 5/d ▪ 5/d = e

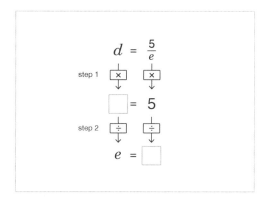

10. Rearrange the formula $p = (q + 3)/7$ to make q the
subject.

a
b
c   ▪ 7p - 3 = q ▪ q = 7p - 3

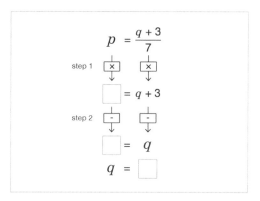

Level 2:   Fluency - Using 2 or more variables with 2 or
more operations.

✸ **Required:** 7/10   ✸ **Student Navigation:** on
✸ **Randomised:** off

11. Rearrange $v = u + at$ to make u the subject of the
formula.

a
b
c   ▪ u = v - at ▪ v - at = u

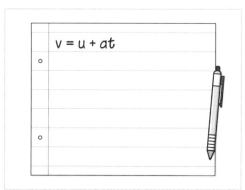

12. Make x the subject of $y = mx + c$.

a
b
c   ▪ x = (y - c)/m

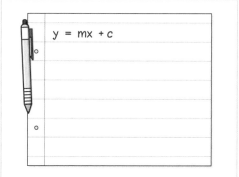

13. Rearrange the formula $p = q/4 - 2r$ to make q the subject of the formula.

a
b
c

- 4(p + 2r) = q ■ q = 4(p + 2r) ■ q = 4p + 8r
- 4p + 8r = q

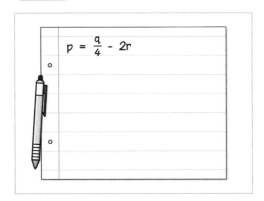

$$p = \frac{q}{4} - 2r$$

14. Make k the subject of the formula $p = -kw$

a
b
c

- k = p/-w ■ k = -p/w

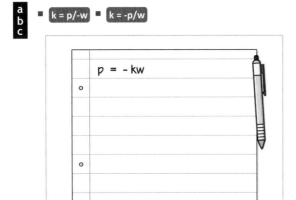

$$p = -kw$$

15. Select the formula which makes h the subject of $S = \pi r^2 + 2\pi rh$.

■ A ■ B ■ C ■ D

1/4

A) $h = S - \pi r^2 - 2\pi r$

B) $h = \dfrac{S}{\pi r^2 - 2\pi r}$

C) $h = \dfrac{S - \pi r^2}{2\pi r}$

D) $h = \dfrac{S}{\pi r^2 + 2\pi r}$

16. Select the formula which makes a the subject of $v^2 = u^2 + 2as$.

■ A ■ B ■ C ■ D

1/4

A) $a = \dfrac{v^2 - \sqrt{u}}{2s}$

B) $a = \dfrac{v^2 - u^2}{2s}$

C) $a = \dfrac{v^2 - u^2}{s}$

D) $a = \dfrac{v^2}{2s} - u^2$

17. Select the **two** correct formulae which make b the subject of $A = (a + b)h/2$.

■ A ■ B ■ C ■ D

2/4

A) $b = 2A - ah$

B) $b = \dfrac{2A}{ah}$

C) $b = \dfrac{2A}{h} - a$

D) $b = \dfrac{2A - ah}{h}$

18. Select the formulae which makes t the subject of $s = 3(t - 2)$.

■ A ■ B ■ C ■ D

2/4

A) $t = \dfrac{s + 2}{3}$

B) $t = \dfrac{s}{3} + 2$

C) $t = \dfrac{s + 6}{3}$

D) $3t = s + 6$

Level 2: cont.

19. Rearrange the formula V = *lbw* to make *w* the subject.

a
b
c
- w=V/lb - w = V/bl

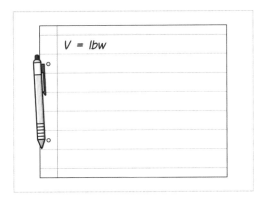

$V = lbw$

20. Select the formula which makes *k* the subject of D = *ut* + *kt²*.

- (i) - (ii) - (iii) - (iv)

1/4

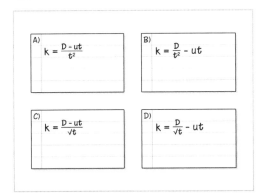

A)
$$k = \frac{D - ut}{t^2}$$

B)
$$k = \frac{D}{t^2} - ut$$

C)
$$k = \frac{D - ut}{\sqrt{t}}$$

D)
$$k = \frac{D}{\sqrt{t}} - ut$$

Level 3: Reasoning - Rearranging formulae.

❋ **Required:** 5/5   ❋ **Student Navigation:** on
❋ **Randomised:** off

21. Joe and Zoe both rearrange a formula to make *d* the subject. Who works out the answer correctly?

1/4

- Joe - Zoe - Both Joe and Zoe - Neither Joe or Zoe

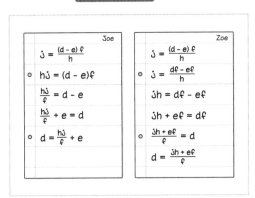

Joe
$$j = \frac{(d - e)f}{h}$$
$$hj = (d - e)f$$
$$\frac{hj}{f} = d - e$$
$$\frac{hj}{f} + e = d$$
$$d = \frac{hj}{f} + e$$

Zoe
$$j = \frac{(d - e)f}{h}$$
$$j = \frac{df - ef}{h}$$
$$jh = df - ef$$
$$jh + ef = df$$
$$\frac{jh + ef}{f} = d$$
$$d = \frac{jh + ef}{f}$$

22. Billy rearranged a formula to make *u* the subject. At which step did he made a mistake?

- Step 1 - Step 2 - Step 3

1/3

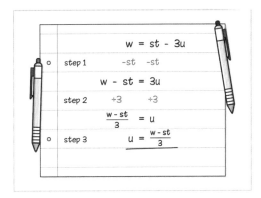

$$w = st - 3u$$
step 1   −st   −st
$$w - st = 3u$$
step 2   ÷3   ÷3
$$\frac{w - st}{3} = u$$
step 3 $$u = \frac{w - st}{3}$$

23. The area of a rectangle is given by the formula A = *lw*.
The area of a triangle is given by the formula A = ½*bh*.
If the area of the two shapes is the same and *l* and *b* are equal, what is the relationship between *w* and *h*?

a
b
c

24. To convert degrees Fahrenheit (F) to degrees Celsius (C) you subtract 32, multiply by 5 and then divide by 9. Write a formula to convert degrees Celsius to degrees Fahrenheit.

a
b
c
- 9/5C + 32 = F - F = 9/5C + 32 - 9C/5 + 32 = F
- F = 9C/5 + 32

25. Select the **two** equivalent formulae.

- (i) - (ii) - (iii) - (iv)

2/4

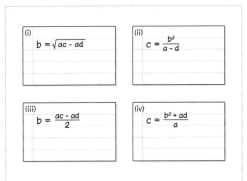

(i)
$$b = \sqrt{ac - ad}$$

(ii)
$$c = \frac{b^2}{a - d}$$

(iii)
$$b = \frac{ac - ad}{2}$$

(iv)
$$c = \frac{b^2 + ad}{a}$$

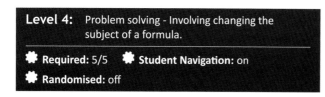

Level 4: Problem solving - Involving changing the subject of a formula.

✻ **Required:** 5/5   ✻ **Student Navigation:** on
✻ **Randomised:** off

26. Molly is competing in a triathlon. The formula
a
b   calculates the distance she has travelled. What
c   formula can she use to calculate her final speed?

- 2S/t - u = v   - (2s-ut)/t = v   - v = (2s-ut)/t
- v = 2S/t - u

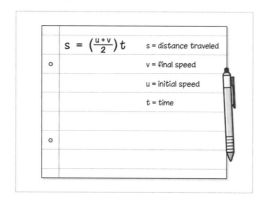

$$s = \left(\frac{u+v}{2}\right)t \qquad s = \text{distance traveled}$$

v = final speed

u = initial speed

t = time

27. A formula is written to calculate the total
☐
☒   perimeter P of 3 regular hexagons with side-length
☐   h and 2 regular decagons with side-length d.
1/5   Rearrange the formula to find the side-length of
the regular decagon. Which option represents
your formula?

- (i)   - (ii)   - (iii)   - (iv)   - (v)

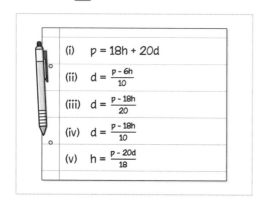

(i)   p = 18h + 20d

(ii) $d = \frac{p - 6h}{10}$

(iii) $d = \frac{p - 18h}{20}$

(iv) $d = \frac{p - 18h}{10}$

(v) $h = \frac{p - 20d}{18}$

28. If A is the shaded area, which formula could you
☐
☒   use to find the value of x?
☐

- (i)   - (ii)   - (iii)   - (iv)

1/4

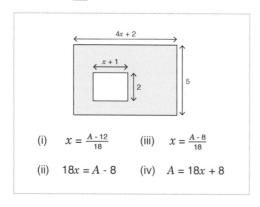

(i) $x = \frac{A - 12}{18}$ (iii) $x = \frac{A - 8}{18}$

(ii) $18x = A - 8$ (iv) $A = 18x + 8$

29. Sam earns e pounds. He has worked x hours at £12
☐   an hour during the week and y hours at £15 an
☒   hour during the weekend. Select the formula you
☐   would use to calculate how many hours he worked
1/4   at the weekend.

- (i)   - (ii)   - (iii)   - (iv)

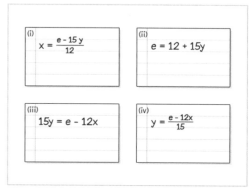

(i) $x = \frac{e - 15y}{12}$ (ii)   e = 12 + 15y

(iii)   15y = e - 12x   (iv) $y = \frac{e - 12x}{15}$

30. Anna thinks of a number y. She adds x to y and
a   then divides the answer by 4. She gets exactly the
b   same answer if she starts with the same number,
c   adds r and divides the answer by 6. Write a
formula in its **simplest form** to find the value of
the number she is thinking of.

- y = 2r - 3x

Substitute Numerical Values (Advanced)

Competency: Substitute numerical values into formulae and expressions, including scientific formulae.

Quick Search Ref: 10081

Correct: Correct.   Wrong: Incorrect, try again.   Open: Thank you.

Level 1: Understanding - Understand how to substitute numerical values to evaluate formulas and expressions.

✿ **Required:** 7/10   ✿ **Student Navigation:** on   ✿ **Randomised:** off

---

1. Give the value for the expression when:
$a = 10$ and $b = 2$.

1
2
3 ▪ 5

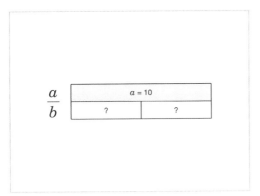

2. Give the value for the expression when:
$s = 11$ and $t = 4$.

1
2
3 ▪ 8.25

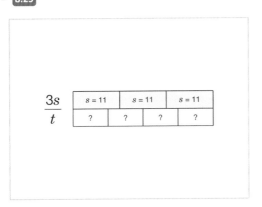

3. Evaluate the expression when:
$s = 2$, $t = 4$, $u = 4$ and $v = 0.8$.

1
2
3 ▪ 2.8

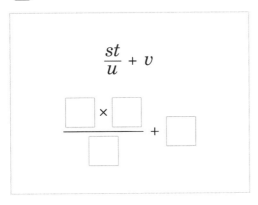

4. Evaluate the expression when:
$a = 21$, $b = 5$ and $c = 3$.

1
2
3 ▪ 12

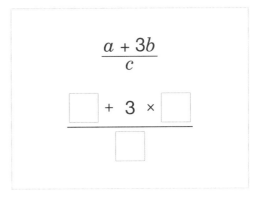

5. Evaluate the expression when:
$a = 13$, $b = 7$, and $c = ½$.

1
2
3 ▪ 0.95

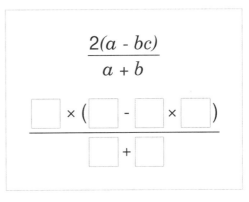

6. Evaluate the expression when:
$n = 3$ and $m = 0.5$.

1
2
3 ▪ 90

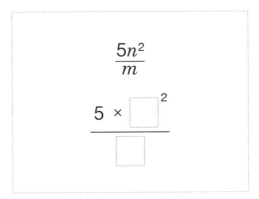

Level 1: cont.

7. Evaluate the expression when:
1 2 3 $a = 8$, $c = 40$ and $d = 2$.

▪ 7.5

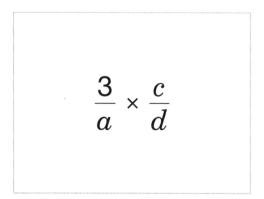

$$\frac{3}{a} \times \frac{c}{d}$$

8. Evaluate the expression when:
1 2 3 $d = 8$, $e = 3$ and $f = 4$.
Give your answer as a decimal.

▪ 1.5

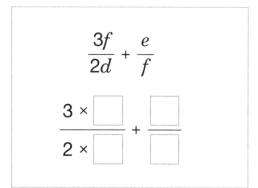

$$\frac{3f}{2d} + \frac{e}{f}$$

$$\frac{3 \times \boxed{}}{2 \times \boxed{}} + \frac{\boxed{}}{\boxed{}}$$

9. Evaluate the expression when:
1 2 3 $e = 3$ and $f = \frac{1}{2}$.

▪ 100

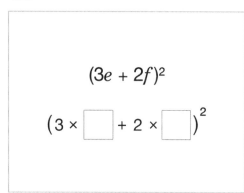

$$(3e + 2f)^2$$

$$\left(3 \times \boxed{} + 2 \times \boxed{}\right)^2$$

10. Evaluate the expression when:
1 2 3 $x = 2$, $y = 5$ and $z = 8$.
Give your answer as a decimal.

▪ -0.5

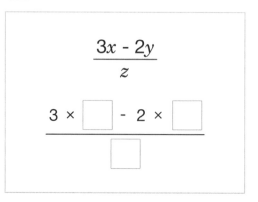

$$\frac{3x - 2y}{z}$$

$$\frac{3 \times \boxed{} - 2 \times \boxed{}}{\boxed{}}$$

Level 2: Fluency - Substitute numerical values including negatives and squares into more complex formulas and expressions.

✿ **Required:** 7/10   ✿ **Student Navigation:** on
✿ **Randomised:** off

11. Which of the cards gives a value of -25 if $n = -3$?

▪ **A** ▪ **B** ▪ **C** ▪ **D** ▪ **E** ▪ **F** ▪ **G**

1/7

(a) $4n$

(b) $-4n$

(c) $3n^2$

(d) $\frac{3n}{2}$

(e) $5n - 10$

(f) $\frac{-6}{n}$

(g) $3n + 8$

12. Evaluate the expression when:
1 2 3 $a = 15$ and $b = -2$.

▪ 25

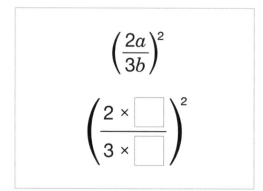

$$\left(\frac{2a}{3b}\right)^2$$

$$\left(\frac{2 \times \boxed{}}{3 \times \boxed{}}\right)^2$$

13. Which of the cards is equal to 0.48 if $n = 0.4$?

▪ A ▪ B ▪ C ▪ D ▪ E ▪ F ▪ G

1/7

(a) $4n$ (b) $-4n$

(c) $3n^2$ (d) $\frac{3n}{2}$

(e) $5n - 10$ (f) $\frac{-6}{n}$

(g) $3n + 8$

14. Evaluate the expression when:
$m = -45$ and $n = -5$.

 ▪ 4

$$\frac{10 - 2m}{n^2}$$

$$\frac{10 - 2 \times \boxed{}}{\boxed{}^2}$$

15. Kinetic energy is given by the formula:
$k = \frac{1}{2}mv^2$. Calculate k when $m = 8$ and $v = 5$.

 ▪ 100

$$k = \frac{1}{2}mv^2$$

$$k = \frac{1}{2} \times \boxed{} \times \boxed{}^2$$

16. Evaluate acceleration using the formula:
$a = (v - u)/t$, when $v = 100$, $u = 22$ and $t = 5$.

 ▪ 15.6

$$a = \frac{(v - u)}{t}$$

$$a = \frac{(\boxed{} - \boxed{})}{\boxed{}}$$

17. Evaluate s (the distance moved) if:
$u = 3.8$ m/s, $t = 2.5$ s and $a = 5.9$ m/s².
Give your answer to two decimal places (2 d.p.).

 ▪ 27.94

u = initial velocity t = time a = acceleration

$$s = ut + \frac{1}{2}at^2$$

$$s = \boxed{} \times \boxed{} + \frac{1}{2} \times \boxed{} \times \boxed{}^2$$

18. The area (A) of a trapezium is given by the formula, where a and b are the lengths of the parallel sides and h is the height of the trapezium. Find the area of trapezium T.
Don't include the units in your answer.

▪ 24

example

$$A = \frac{(a + b)h}{2}$$

trapezium T 3 m 6 m 5 m

$$A = \frac{(\boxed{} + \boxed{}) \times \boxed{}}{2}$$

Level 2: *cont.*

19. Power is calculated using the formula:
$P = I^2R$, where I is current and R is resistance.
Evaluate P when $I = 9$ and $R = \frac{1}{2}$.

▪ **40.5**

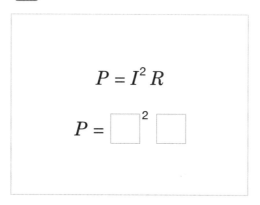

$$P = I^2 R$$

$$P = \boxed{}^2 \boxed{}$$

20. The total surface area (S) of a cuboid is given by the formula where l is the length, h is the height and w is the width.
Find the total surface area S for cuboid C.
Don't include the units in your answer.

▪ **174**

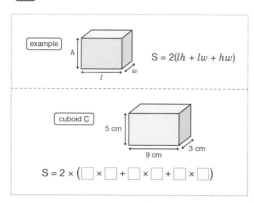

example

$S = 2(lh + lw + hw)$

cuboid C
5 cm
9 cm
3 cm

$S = 2 \times (\boxed{} \times \boxed{} + \boxed{} \times \boxed{} + \boxed{} \times \boxed{})$

Level 3: Reasoning - Substitution into algebraic fractions, checking accuracy and comparing results.

✱ **Required:** 5/5 ✱ **Student Navigation:** on
✱ **Randomised:** off

21. Becky and Andy both have different answers to an algebraic substitution problem. Who is correct? Explain your answer.

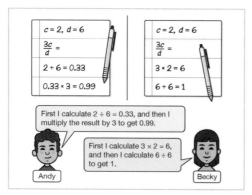

$c = 2, d = 6$	$c = 2, d = 6$
$\frac{3c}{d} =$	$\frac{3c}{d} =$
$2 \div 6 = 0.33$	$3 \times 2 = 6$
$0.33 \times 3 = 0.99$	$6 \div 6 = 1$

First I calculate $2 \div 6 = 0.33$, and then I multiply the result by 3 to get 0.99.

Andy

First I calculate $3 \times 2 = 6$, and then I calculate $6 \div 6$ to get 1.

Becky

22. Sort the expressions in ascending order (smallest value at the top).

↑
↓ ▪ D ▪ A ▪ C ▪ E ▪ B

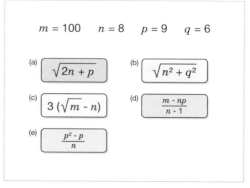

$m = 100 \quad n = 8 \quad p = 9 \quad q = 6$

(a) $\sqrt{2n + p}$

(b) $\sqrt{n^2 + q^2}$

(c) $3(\sqrt{m} - n)$

(d) $\frac{m - np}{n - 1}$

(e) $\frac{p^2 - p}{n}$

23. What number is missing from the following statement:
To make the expression true n must must be a positive number less than __?

▪ **1**

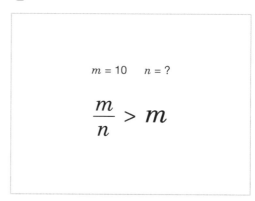

$m = 10 \quad n = ?$

$$\frac{m}{n} > m$$

24. Which one of the following expressions is it not possible to give a value for?

☐
☒
☐

1/7

▪ A ▪ B ▪ C ▪ D ▪ E ▪ F ▪ G

$a = -4 \quad b = -25 \quad c = \frac{1}{4}$

(a) $\sqrt{ab}$

(b) $\sqrt{c}$

(c) $-2b^2$

(d) a^2

(e) $\sqrt{a}$

(f) $3c^2$

(g) $\sqrt{abc}$

25. Which expression is the odd one out? Explain your answer.

a
b
c
Reminder: √100 is the number that when multiplied by itself gives 100.

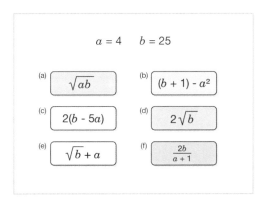

$$a = 4 \quad b = 25$$

(a) $\sqrt{ab}$

(b) $(b + 1) - a^2$

(c) $2(b - 5a)$

(d) $2\sqrt{b}$

(e) $\sqrt{b} + a$

(f) $\dfrac{2b}{a + 1}$

Level 4: Problem Solving - Substitution into algebraic formulae in context and taking further steps to solve a problem.

❇ **Required: 5/5**　❇ **Student Navigation:** on
❇ **Randomised:** off

26. The longest side of a right-angled triangle can be found by using the formula:

1
2
3
$c = \sqrt{(a^2 + b^2)}$
Find the perimeter of square A, when $a = 3$ cm and $b = 4$ cm.
Don't include the units in your answer.

▪ 20

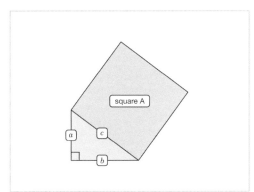

square A

a   c

b

27. Nadia is planting seeds which will germinate at temperatures above 50°F, but her thermometer only has a centigrade scale and also gives inaccurate readings, as much as 3°C above the actual temperature. When taking a reading from her thermometer, at what temperature can Nadia be confident that the seeds will germinate?

1
2
3

▪ 13

To convert degrees fahrenheit to degrees celsius, use the formula:

$$C° = \frac{5(f° - 32)}{9}$$

28. The area of trapezium B is equal to the area of rectangle C. Find the length x for a side of rectangle C.

1
2
3
Don't include the units in your answer.

▪ 3

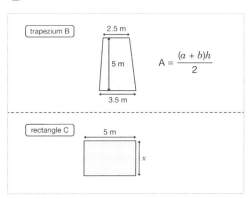

trapezium B

2.5 m

5 m　　$A = \dfrac{(a + b)h}{2}$

3.5 m

rectangle C　　5 m

x

29. The volume of cuboid X is equal to the volume of cube Y.

1
2
3
Find the length y for a side of cube Y when $l = 9$ cm, $w = 4$ cm and $h = 6$ cm.
Don't include the units in your answer.

▪ 6

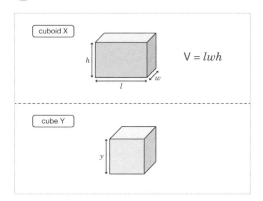

cuboid X

h　　　$V = lwh$

l　　w

cube Y

y

30. Charlie weighs 187 lbs and is 6 feet 2 inches tall (6′ 2″). What is his BMI in kg/m²?
Round your answer to 2 decimal places (2 d.p.) and don't include the units in your answer.
Hint: You need to convert to metric units before you can use the formula.

▪ 24.06

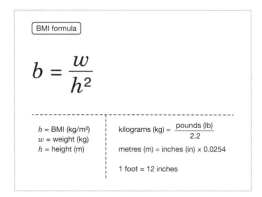

Expand and Simplify Linear Expressions

Competency: Simplify and manipulate algebraic expressions to maintain equivalence by: collecting like terms, multiplying a single term over a bracket, taking out common factors, expanding products of two or more binomials.

Quick Search Ref: 10068

Correct: Correct.   Wrong: Incorrect, try again.   Open: Thank you.

Level 1: Understanding - Expand a single bracket and collect like terms.

✿ **Required:** 7/10   ✿ **Student Navigation:** on   ✿ **Randomised:** off

---

1. Simplify the following expression by collecting like terms:
a
b $4a + 5 + 2a - 3$
c

▪ 6a + 2 ▪ 2 + 6a

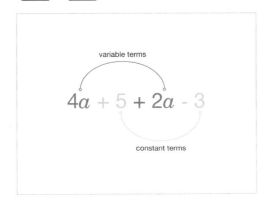

2. Expand the following:
a
b $7(3x + 8)$
c

▪ 21x + 56 ▪ 56 + 21x

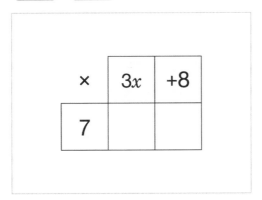

3. Multiply out $3(6e - 5)$.
a
b ▪ 18e - 15 ▪ -15 + 18e
c

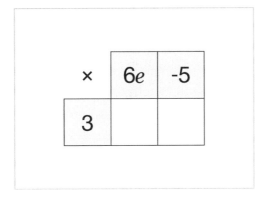

4. Expand and simplify:
a
b $2(4y + 7) + 5$
c ▪ 8y + 19 ▪ 19 + 8y

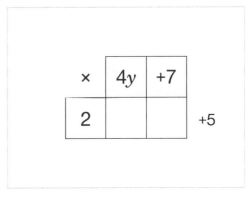

5. Multiply out the bracket and simplify:
a
b $3c + 8(2c - 1)$
c ▪ -8 + 19c ▪ 19c - 8

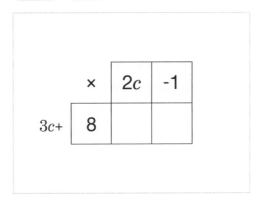

6. Expand and simplify the following:
a
b $6(4r - 3) + 20$
c ▪ 2 + 24r ▪ 24r + 2

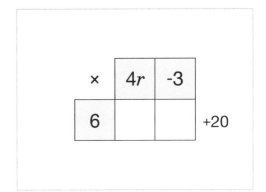

Level 1: cont.

7. Multiply and simplify:
a
b
c
$4(5 - 9t) + 20t$
▪ -16t + 20 ▪ 20 - 16t

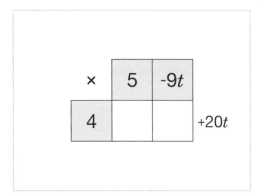

8. Expand and simplify:
a
b
c
$2(7s + 4) + 6$
▪ 14s + 14 ▪ 14 + 14s

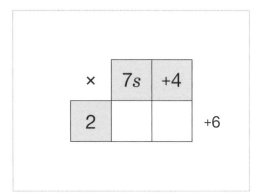

9. Expand and simplify:
a
b
c
$5(8a - 5) - 9a$
▪ -25 + 31a ▪ 31a - 25

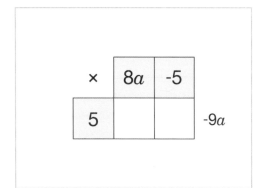

10. Expand and simplify:
a
b
c
$11 + 6(8v - 3)$
▪ -7 + 48v ▪ 48v - 7

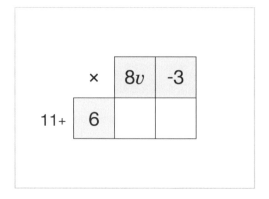

Level 2: Fluency - Expand two brackets and collect like terms.

✻ **Required:** 7/10 ✻ **Student Navigation:** on
✻ **Randomised:** off

11. Expand the brackets and simplify:
a
b
c
$2(4x + 7) + 3(5x + 6)$
▪ 23x + 32 ▪ 32 + 23x

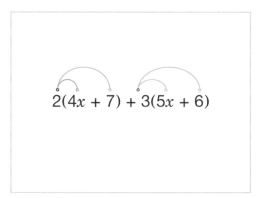

12. Expand and simplify the following expression:
a
b
c
$4(6b - 1) + 2(7b + 3)$
▪ 2 + 38b ▪ 38b + 2

$$4(6b - 1) + 2(7b + 3)$$

Level 2: cont.

13. Expand and simplify:

$5(3t + 4) + 8(2t - 7)$

- -36 + 31t
- 31t - 36

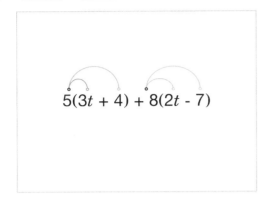

14. Expand and simplify the following expression:

$7(4x + 9) - 2(3x + 2)$

- 59 + 22x
- 22x + 59

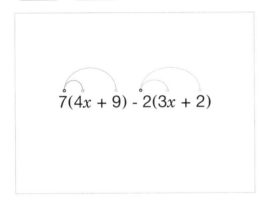

15. Expand and simplify:

$12(6g + 7) - 5(8g - 3)$

- 99 + 32g
- 32g + 99

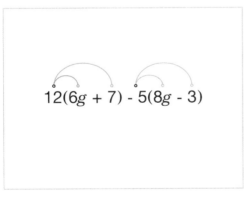

16. Expand and simplify:

$4(z - 6) + 7(9z - 2)$

- 67z - 38
- -38 + 67z

17. Expand and simplify:

$8(3a + 11) - (5a - 12)$

- 100 + 19a
- 19a + 100

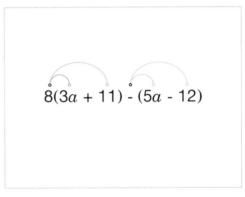

18. Expand and simplify the following expression:

$6(2c + 5) + 9(4c - 7)$

- -33 + 48c
- 48c - 33

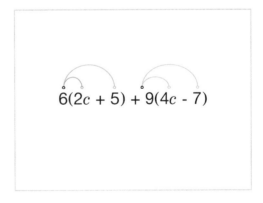

19. Expand and simplify:

a
b
c
$10(2w + 11) - 6(5w - 8)$

▪ 158 - 10w ▪ -10w + 158

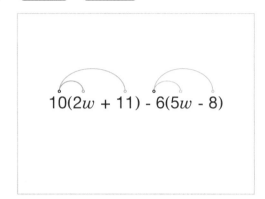

$$10(2w + 11) - 6(5w - 8)$$

20. Expand and simplify:

a
b
c
$3(5r + 12) - (4r - 9)$

▪ 11r + 45 ▪ 45 + 11r

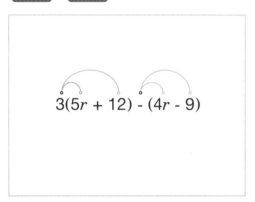

$$3(5r + 12) - (4r - 9)$$

Level 3: Reasoning - Inverse questions and recognising mistakes with expanding and simplifying.

✱ **Required:** 5/5 ✱ **Student Navigation:** on
✱ **Randomised:** off

21. In the expression shown, what is the missing term?

a
b
c
▪ 4x

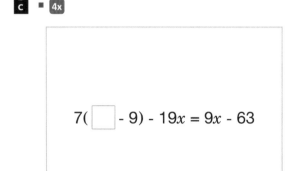

$$7(\boxed{} - 9) - 19x = 9x - 63$$

22. Expand and simplify the following expressions and arrange in ascending order (smallest first).

↑
↓
▪ 3(4x + 1) + 6x ▪ 6(8x + 3) - 2(15x + 7) ▪ 2(9x + 1) + 3

23. Zoe has made one mistake in her homework on expanding and simplifying expressions.

a
b
c
What is the correct answer to the question she got wrong?

▪ 18c + 39

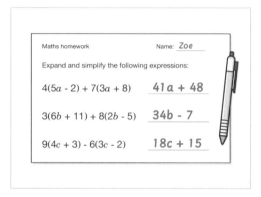

Maths homework   Name: *Zoe*

Expand and simplify the following expressions:

$4(5a - 2) + 7(3a + 8)$ **41a + 48**

$3(6b + 11) + 8(2b - 5)$ **34b - 7**

$9(4c + 3) - 6(3c - 2)$ **18c + 15**

24. Tara says that

a
b
c
$14x - 5(7 - 2x)$ is equal to $14x + 5(2x - 7)$.
Is Tara correct? Explain your answer.

25. Sammy has made a mistake expanding and simplifying the expression shown.

a
b
c
What is the correct simplification?

▪ 1x + 26 ▪ 26 + x ▪ 26 + 1x ▪ x + 26

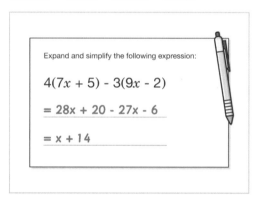

Expand and simplify the following expression:

$4(7x + 5) - 3(9x - 2)$

$= 28x + 20 - 27x - 6$

$= x + 14$

Level 4: Problem Solving - Multi-step problems expanding and simplifying expressions.

✱ **Required:** 5/5 ✱ **Student Navigation:** on
✱ **Randomised:** off

26. What is the area of the shape?

a
b
c
Simplify your answer.

▪ 86x -16 ▪ -16 + 86x

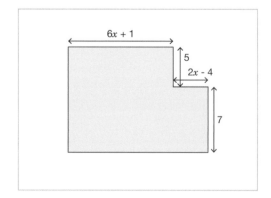

$6x + 1$

5

$2x - 4$

7

27. Find the value of *x* if 7(3*x* - 2) - 5*x* = 42.
Give your answer as a decimal.

1
2
3

▪ 3.5

28. Write an expression for the surface area of the cuboid.

a
b
c

▪ 12 + 66x ▪ 66x + 12

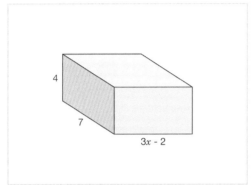

4

7

3*x* - 2

29. The grid contains 4 pairs of equivalent expressions. Expand and simplify the odd one out.

a
b
c

▪ 10x - 63 ▪ -63 + 10x

8(3*x* + 1) - 5*x*	28*x* - 3	8(2*x* - 3) + 3*x*
19*x* - 24	3(12*x* - 1) + 7	4(7*x* + 2) - 11
7(2*x* - 9) - 4*x*	19*x* + 8	36*x* + 4

30. A sequence has a term-to-term rule of *double the previous term and then add 3*. If the sum of the first four terms is 108, what is the fourth term in the sequence?

1
2
3

▪ 61

Expand the Product of Three or More Binomials

Competency: Simplify and manipulate algebraic expressions to maintain equivalence by: collecting like terms; multiplying a single term over a bracket; taking out common factors; expanding products of two or more binomials.

Quick Search Ref: 10276

Correct: Correct.   **Wrong:** Incorrect, try again.   **Open:** Thank you.

Level 1: Understanding - Expand the product of a binomial and a trinomial.

✱ **Required:** 4/5   ✱ **Student Navigation:** on   ✱ **Randomised:** off

---

1. What expression completes the expansion of
$(x + 4)(x^2 + 5x + 6)$?

a
b
c
  ▪ 26x + 24   ▪ 24 + 26x

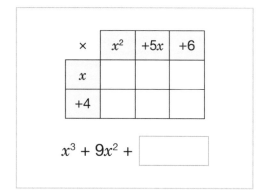

2. Expand and simplify
$(2x + 5)(x^2 + 7x - 1)$.

1/7

▪ $2x^3 + 19x^2 + 33x + 5$ ▪ $2x^3 + 19x^2 + 35x - 5$
▪ $2x^3 + 5x^2 + 35x - 5$ ▪ $2x^3 + 19x^2 + 35x + 5$
▪ $2x^3 + 5x^2 + 33x - 5$ ▪ $2x^3 + 19x^2 + 33x - 5$
▪ $2x^3 + 5x^2 + 33x + 5$

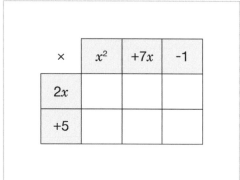

3. Select the four terms that make up the expansion of
$(5x - 2)(2x^2 + 6x + 9)$.

4/7   ▪ +18   ▪ +26x²   ▪ -4x²   ▪ 33x   ▪ -12x   ▪ 10x³   ▪ -18

4. Expand and simplify
$(7x^2 + 8x - 2)(3x - 5)$.

1/6

▪ $21x^3 - 35x^2 - 40x + 10$ ▪ $21x^3 - 11x^2 - 40x + 10$
▪ $21x^3 - 35x^2 - 6x + 10$ ▪ $21x^3 - 11x^2 - 46x + 10$
▪ $21x^3 - 35x^2 - 46x + 10$ ▪ $21x^3 - 11x^2 - 6x + 10$

5. Select the terms that make up the expansion of
$(3x^2 - 8x + 4)(2x - 9)$.

4/7   ▪ 6x³   ▪ -27x²   ▪ +80x   ▪ -36   ▪ +36   ▪ -43x²   ▪ 8x

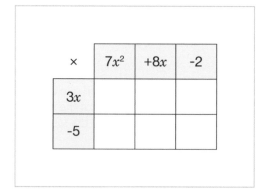

Level 2: Fluency - Expand the product of three binomials.

✱ **Required:** 4/5   ✱ **Student Navigation:** on
✱ **Randomised:** off

6. Select the four terms that make up the expansion of
$(x + 4)(x + 7)(x + 3)$.

4/7 ▪ x^3 ▪ $11x^2$ ▪ $18x^2$ ▪ $14x^2$ ▪ $61x$ ▪ $28x$ ▪ 84

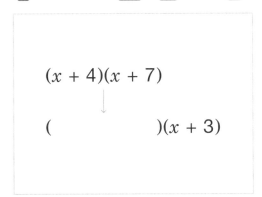

7. Expand and simplify
$(2x + 9)(x + 3)(5x + 2)$.

1/6 ▪ $10x^3 + 79x^2 + 30x + 54$ ▪ $10x^3 + 4x^2 + 135x + 54$
▪ $10x^3 + 79x^2 + 135x + 54$ ▪ $10x^3 + 4x^2 + 30x + 54$
▪ $10x^3 + 79x^2 + 165x + 54$ ▪ $10x^3 + 4x^2 + 165x + 54$

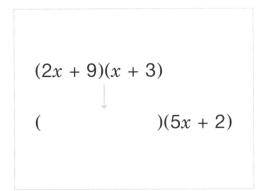

8. What expression completes the expansion of
$(4x - 7)(9x + 1)(3x - 2)$?

▪ $14 + 97x$ ▪ $97x + 14$

9. Expand and simplify
$(2x - 3)(5x + 2)(6x - 7)$.

1/6 ▪ $60x^3 - 70x^2 - 36x + 42$ ▪ $60x^3 - 136x^2 - 36x + 42$
▪ $60x^3 - 70x^2 + 41x + 42$ ▪ $60x^3 - 136x^2 + 41x + 42$
▪ $60x^3 - 70x^2 + 77x + 42$ ▪ $60x^3 - 136x^2 + 77x + 42$

10. Select the three terms that make up the expansion of
$(6x + 1)(2x + 5)(3x - 8)$.

3/7 ▪ $-12x^2$ ▪ $36x^3$ ▪ $-241x$ ▪ $-185x$ ▪ -40 ▪ $+12x^2$
▪ $+11x$

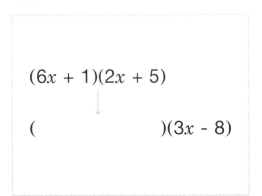

Level 3: Reasoning - Finding coefficients and terms when using product of binomials.

✱ **Required:** 3/3   ✱ **Student Navigation:** on
✱ **Randomised:** off

11. How many terms are there when expanding
$(x + a)(x + b)(x + c)$?

1
2
3
▪ 8

12. Jayden has spilt ink on his maths notes. What expression is underneath the ink spot?

▪ $3x - 2$ ▪ $(3x - 2)$ ▪ $(-2 + 3x)$ ▪ $-2 + 3x$

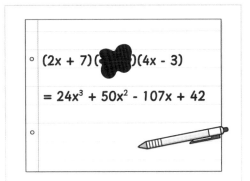

Level 3: cont.

13. When $(x + 1)^4$ is expanded and simplified, what is the coefficient of the x^2 term?

1
2
3

▪ 6

Level 4: Problem Solving - Multi-step problems using products of binomials.

✱ **Required:** 3/3   ✱ **Student Navigation:** on
✱ **Randomised:** off

14. What expression completes the calculation to find the volume of the cuboid?

a
b
c

▪ + 30 - 19x   ▪ 30 - 19x   ▪ - 19x + 30

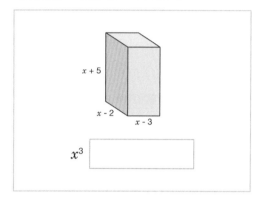

x^3 [   ]

15. A cube has side lengths of x metres. A cuboid has side lengths of x, $(x + 1)$ and $(x - 1)$ metres. What expression represents the difference in volume of the two solids?

a
b
c

▪ x   ▪ -x

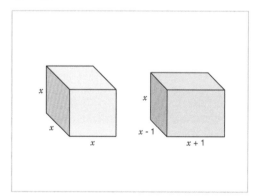

16. A cuboid has a square cross-section with side length $2x + 1$. If the volume of the cuboid is $12x^3 + 8x^2 - x - 1$, what expression represents the length of the cuboid?

a
b
c

▪ -1 + 3x   ▪ 3x - 1

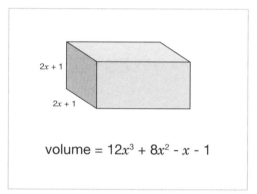

volume = $12x^3 + 8x^2 - x - 1$

Expand the Product of Two Binomials

Competency: Simplify and manipulate algebraic expressions to maintain equivalence by: collecting like terms; multiplying a single term over a bracket; taking out common factors; expanding products of two or more binomials.

Quick Search Ref: 10298

Correct: Correct.   Wrong: Incorrect, try again.   Open: Thank you.

Level 1: Understanding - Expanding binomials with positive terms and variables with coefficients of 1.

✱ **Required:** 7/10   ✱ **Student Navigation:** on   ✱ **Randomised:** off

---

1. What constant term completes the expansion of $(x + 2)(x + 3)$?

- 6

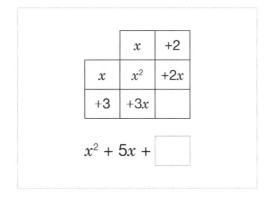

2. Expand and simplify $(a + 5)(a + 7)$.

- $a^2 + 12a + 12$ ■ $14a + 35$ ■ $a^2 + 35$ ■ $12a + 35$
- $a^2 + 12a + 35$ ■ $14a + 12$ ■ $a^2 + 5a + 7a + 35$

1/7

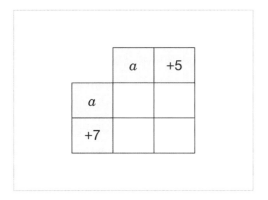

3. Select the three terms that make up the expansion of $(x + 8)(x + 4)$.

3/7   ■ x^2 ■ $2x^2$ ■ $12x$ ■ $14x$ ■ $32x$ ■ 12 ■ 32

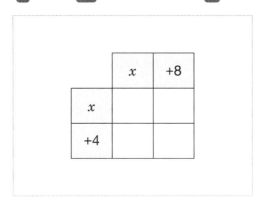

4. Expand and simplify $(y + 3)(y + 3)$.

- $y^2 + 6y + 6$ ■ $y^2 + 6$ ■ $y^2 + 6y + 9$ ■ $y^2 + 9$ ■ $y^2 + 9y + 6$
- $y^2 + 3$ ■ $y^2 + 9y + 9$

1/7

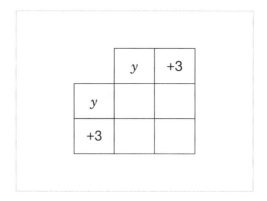

5. What term completes the expansion of $(x + 7)(x + 12)$?

- $+19x$ ■ $19x$

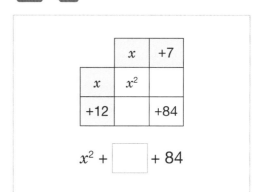

6. Select the three terms that make up the expansion of $(t + 1)^2$.

- $2t^2$ ■ 1 ■ t ■ t^2 ■ 2 ■ $2t$

3/6

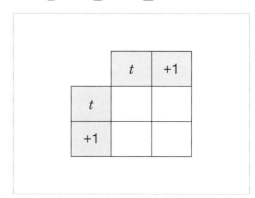

Level 1: *cont.*

7. What expression completes the expansion of

$(x + 9)(x + 8)$?

■ 72 + 17x ■ 17x + 72

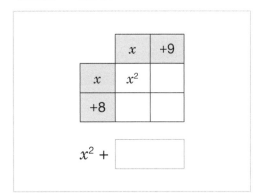

$x^2 +$ ☐

8. Expand and simplify $(v + 3)(v + 5)$.

■ v² + 15v + 8 ■ v² + 15v + 15 ■ v² + 8v + 15
■ v² + 3v + 5v + 15 ■ v² + 8v + 8 ■ v² + 15 ■ v² + 8

1/7

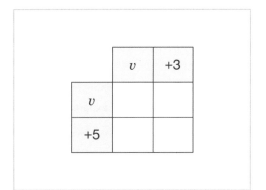

9. Select the three terms that make up the expansion of $(x + 4)^2$.

■ 8 ■ 8x ■ 4 ■ x² ■ 2x ■ 2x² ■ 16

3/7

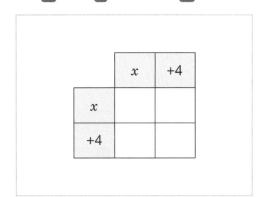

10. What expression completes the expansion of $(m + 11)(m + 9)$?

■ 20m + 99 ■ 99 + 20m

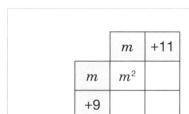

$m^2 +$ ☐

Level 2: Fluency - Expanding binomials with negative terms and variables with coefficients greater than 1.

✹ **Required:** 7/10 ✹ **Student Navigation:** on
✹ **Randomised:** off

11. Expand and simplify $(2x + 1)(x + 6)$.

■ 2x² + 7x + 6 ■ x² + 13x + 6 ■ 2x² + 13x + 7 ■ x² + 7x + 6
■ 2x² + 13x + 6 ■ x² + 7x + 7

1/6

$$(2x + 1)(x + 6)$$

12. Select the three terms that make up the expansion of $(5c + 4)(3c + 2)$.

■ +8c² ■ +15c² ■ +6c ■ +8c ■ +22c ■ +6 ■ +8

3/7

$$(5c + 4)(3c + 2)$$

Level 2: cont.

13. What term completes the expansion of

 $(x + 7)(x - 2)$?

■ 5x ■ +5x

$$(x + 7)(x - 2)$$

$$x^2 + \boxed{} - 14$$

14. Expand and simplify

$(r - 5)(3r + 8)$.

■ $3r^2 + 23r - 40$ ■ $r^2 - 7r + 3$ ■ $r^2 + 23r - 13$ ■ $3r^2 - 7r - 40$

1/6 ■ $r^2 + 7r - 40$ ■ $3r^2 - 23r + 3$

$$(r - 5)(3r + 8)$$

15. Select the terms that make up the expansion of

 $(4x + 7)(6x - 3)$.

 ■ -21 ■ +54x ■ $10x^2$ ■ +4 ■ +30x ■ $24x^2$ ■ +21

3/7

$$(4x + 7)(6x - 3)$$

16. What expression completes the expansion of

$(9n - 5)(7n + 8)$?

■ + 37n - 40 ■ 37n - 40 ■ -40 + 37n

$$(9n - 5)(7n + 8)$$

$$63n^2 + \boxed{}$$

17. Select the terms that make up the expansion of

$(5x - 3)(2x - 9)$.

■ +27 ■ -51x ■ $7x^2$ ■ -27 ■ $10x^2$ ■ -12 ■ +51x

3/7

$$(5x - 3)(2x - 9)$$

18. Expand and simplify

$(6w + 7)(8w + 3)$.

■ $48w^2 + 18w + 10$ ■ $14w^2 + 56w + 10$ ■ $48w^2 + 74w + 21$

1/6 ■ $14w^2 + 74w + 10$ ■ $48w^2 + 56w + 21$ ■ $14w^2 + 18w + 21$

$$(6w + 7)(8w + 3)$$

Level 2: *cont.*

19. Select the terms that make up the expansion of (4x - 3)(9x - 1).

☐ ☒ ☐

■ -3 ■ 36x² ■ -31x ■ 13x² ■ +3 ■ -4 ■ +31x

3/7

$$(4x - 3)(9x - 1)$$

20. What expression completes the expansion of (7k - 4)(8k + 5)?

a b c

■ -20 + 3k ■ 3k - 20

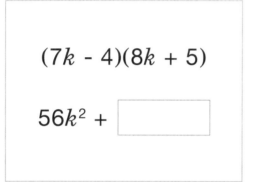

$$(7k - 4)(8k + 5)$$

$$56k^2 + \boxed{}$$

Level 3: Reasoning - Misconceptions and the difference of two squares.

✹ **Required:** 5/5   ✹ **Student Navigation:** on
✹ **Randomised:** off

21. How many terms are there when you expand and simplify (x + a)(x - a)?

1 2 3

■ 2

22. What missing expression makes the following statement true?

a b c

(x -2)(?) = 2x² - x - 6.

■ (3 + 2x) ■ (2x + 3) ■ 3 + 2x ■ 2x + 3

$$(x - 2)(\boxed{}) = 2x^2 - x - 6$$

23. Jenny says that (x + 5)² = x² + 5² = x² + 25.

a b c

Is Jenny correct? Explain your answer.

24. Without using a calculator or formal written methods find the value of 801 × 799.

a b c

Hint: 801 × 799 = (800 + 1)(800 - 1).

■ 639,999 ■ 639999

25. A piece of string is arranged to make a square with a side length of *x* metres.

1 2 3

The string is then rearranged to make a rectangle with one pair of sides (x + 1) metres and the other pair (x - 1) metres.

What is the difference in square metres between the area of the square and the rectangle?

Do not include units in your answer.

■ 1 ■ -1

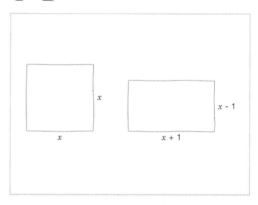

Level 4: Problem Solving - Multi-step problems expanding binomials.

✹ **Required:** 5/5   ✹ **Student Navigation:** on
✹ **Randomised:** off

26. Complete the expression for the area of the triangle.

a b c

■ 1x - 1 ■ -1 + 1x ■ -1 + x ■ x - 1

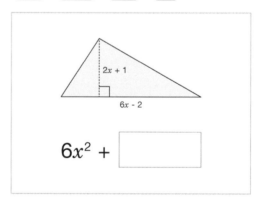

$$6x^2 + \boxed{}$$

27. A square has a side length of *x* metres.

a b c

If the side length is increased by 3 metres, what is the expression for the increase in area of the square in metres squared?

Do not include units in your answer.

■ 6x + 9 ■ 9 + 6x

28. The grid shows 9 expressions.

 Four pairs of expressions are equivalent.
Expand and simplify all the expressions to find the odd one out.

Give the *x* term of the odd one out.

- **22x**

$(2x + 1)(6x - 5)$	$12x^2 - 17x + 6$	$12x^2 + 5x - 2$
$(4x - 1)(3x + 2)$	$12x^2 - 20x + 3$	$(3x - 2)(4x - 3)$
$(6x + 2)(2x + 3)$	$12x^2 - 4x - 5$	$(6x - 1)(2x - 3)$

29. Write an expression for the area of the shape.

 - **22x + 68** - **68 + 22x**

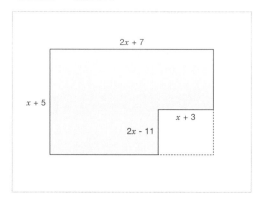

30. The triangle and the square have the same area.
 What is the value of *x*?

- **11**

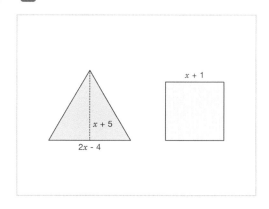

Factorise expressions using brackets

Competency: Simplify and manipulate algebraic expressions to maintain equivalence by taking out common factors.

Quick Search Ref: 10111

Correct: Correct.   **Wrong:** Incorrect, try again.   **Open:** Thank you.

Level 1: Understanding - Factorising expressions with a constant HCF.

❇ Required: 7/10   ❇ Student Navigation: on   ❇ Randomised: off

---

1. Which is the best description of factorising?

- Multiplying an expression by itself.
- Writing an expression as a product of its factors.

1/3
- Multiplying every term in a bracket by the term outside the bracket.

2. You can check an expression is correct by expanding it. Expand the following expression: $7(5m + 9)$

a
b
c

- 35m + 63   ▪ 63 + 35m

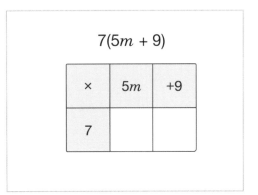

3. What missing number completes the factorisation of $10x + 15$?

 ▪ 3

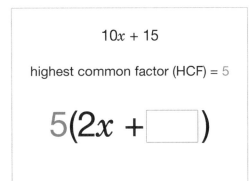

4. What missing number completes the factorisation of $21a - 14b$?

1
2
3
▪ 7

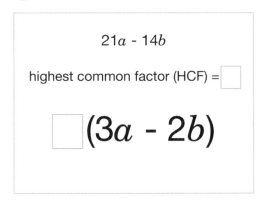

$21a - 14b$

highest common factor (HCF) = ▢

▢$(3a - 2b)$

5. Which expression is the full factorisation of $36r + 27$?

- 9(4r + 27)   ▪ 3(12r + 9)   ▪ 9(4r + 18)   ▪ 9(4r + 3)

1/4

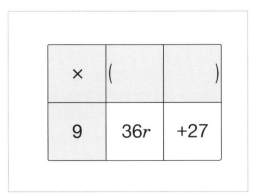

6. Factorise $45v - 20$.

a
b
c

▪ 5(-4 + 9v)   ▪ 5(9v - 4)

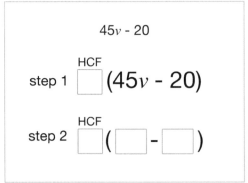

$45v - 20$

step 1 ▢(HCF)$(45v - 20)$

step 2 ▢(HCF)$(▢ - ▢)$

                   Ref:10111   Factorise expressions using brackets

Level 1: cont.

7. Fully factorise $42m + 36n$.

a
b
c

■ 6(7m + 6n)   ■ 6(6n + 7m)

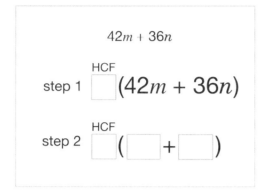

$$42m + 36n$$

step 1 $\overset{\text{HCF}}{\Box}(42m + 36n)$

step 2 $\overset{\text{HCF}}{\Box}(\Box + \Box)$

8. Which expression is the full factorisation of $78c - 54d$?

1/5

■ 2(39c - 27d) ■ 6(13c - 54d) ■ 3(26c - 18d) ■ 6(13c - 9d)
■ 2(39c - 54d)

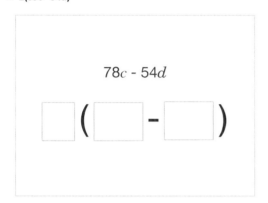

$$78c - 54d$$

$$\Box (\Box - \Box)$$

9. Fully factorise $56k + 70$.

a
b
c

■ 14(4k + 5)   ■ 14(5 + 4k)

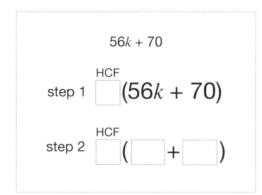

$$56k + 70$$

step 1 $\overset{\text{HCF}}{\Box}(56k + 70)$

step 2 $\overset{\text{HCF}}{\Box}(\Box + \Box)$

10. Fully factorise $60p - 75q$.

a
b
c

■ 15(-5q + 4p)   ■ 15(4p - 5q)

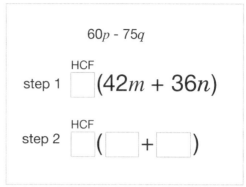

$$60p - 75q$$

step 1 $\overset{\text{HCF}}{\Box}(42m + 36n)$

step 2 $\overset{\text{HCF}}{\Box}(\Box + \Box)$

Level 2:   Fluency - Factorising expressions with a HCF containing variables.

✸ **Required:** 7/10   ✸ **Student Navigation:** on
✸ **Randomised:** off

11. What is the highest common factor of $6ab$ and $8a$?

a
b
c

■ 2a

12. What missing term completes the factorisation of $10m^2 + 8m$?

a
b
c

■ 5m

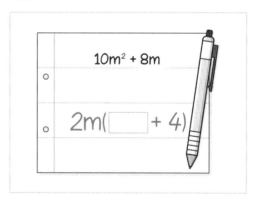

$$10m^2 + 8m$$

$$2m(\Box + 4)$$

13. What missing term completes the factorisation of $24r^2 - 15rs$?

a
b
c

■ 3r

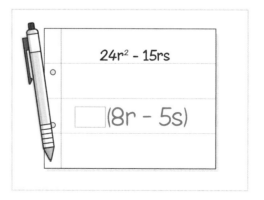

$$24r^2 - 15rs$$

$$\Box(8r - 5s)$$

Level 2: *cont.*

14. Fully factorise 16*xy* + 12*xz*.

■ 4x(4y + 3z)   ■ 4x(3z + 4y)

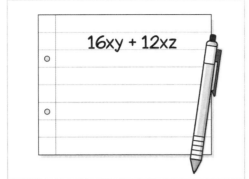

$$16xy + 12xz$$

15. What is the full factorisation of 42*u²v* - 28*uv²*?

■ 14uv(-2v + 3u) ■ 14vu(3u - 2v) ■ 14vu(-2v + 3u)
■ 14uv(3u - 2v)

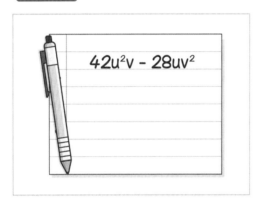

$$42u^2v - 28uv^2$$

16. Which is the correct full factorisation of 24*d³e* + 54*d²e²*?

■ 6de(4d² + 9de) ■ 6d²e(4d + 9e) ■ 6d²(4de + 9e²)

1/4   ■ 6d²e²(4d + 9)

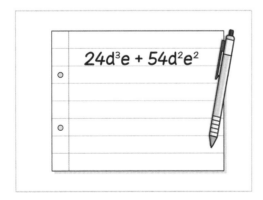

$$24d^3e + 54d^2e^2$$

17. Fully factorise 18*wx* + 12*wy* + 15*wz*.

■ 3w(5z + 6x + 4y)   ■ 3w(4y + 6x + 5z)
■ 3w(4y + 5z + 6x)   ■ 3w(5z + 4y + 6x)
■ 3w(6x + 4y + 5z)   ■ 3w(6x + 5z + 4y)

18. Fully factorise 63*m²n* - 54*mn³*.

■ 9mn(7m - 6n²) ■ 9m(7mn - 6n³) ■ 3mn(21m - 18n²)
■ 9n(7m² - 6mn²)

1/4

$$63m^2n - 54mn^3$$

19. Which is the correct full factorisation of 49*rs³t* + 56*s²t²*?

■ 7s²t(7rs + 8t) ■ 7st(7rs² + 8st) ■ 7s²(7rst + 56t²)

1/4   ■ 7s²t²(7rs + 8)

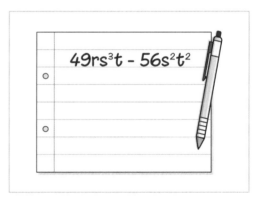

$$49rs^3t - 56s^2t^2$$

20. Factorise 60*hi* + 50*ij* + 15*ik*.

■ 5i(3k + 10j + 12h)   ■ 5i(10j + 12h + 3k)
■ 5i(3k + 12h + 10j)   ■ 5i(12h + 3k + 10j)
■ 5i(10j + 3k + 12h)   ■ 5i(12h + 10j + 3k)

Level 3: Reasoning - Misconceptions and using factorising to solve problems.

✱ **Required:** 5/5   ✱ **Student Navigation:** on
✱ **Randomised:** off

21. Explain why the expression 3*a*(6*a* - 10) is not fully factorised.

22. Without using a calculator, find the value of the following expression:
2.7 × 5.6 - 2.7 × 3.6

■ 5.4

Level 3: *cont.*

23. Maisie made one mistake in her factorising
a homework. What is the correct answer to the
b question she got wrong?
c

▪ 6u(7v + 4u) ▪ 6u(4u + 7v)

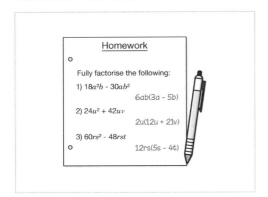

24. Calvin says that 4(3x - 7) is equal to -4(7 - 3x).
a Is Calvin correct? Explain your answer.
b
c

25. If $n^2 + n = 420$ and n is positive, what is the value
1 of n?
2
3 ▪ 20

Level 4: Problem Solving - Multi-step problems
involving factorising.

✹ **Required:** 5/5   ✹ **Student Navigation:** on
✹ **Randomised:** off

26. There are **four pairs** of equivalent expressions in
a the grid. Which expression is the odd one out?
b Factorise your answer.
c

▪ 6ba(-3b + 2a) ▪ 6ab(2a - 3b) ▪ 6ab(-3b + 2a)
▪ 6ba(2a - 3b)

$18ab - 12b^2$	$4ab(3a + 4b)$	$2a(9ab - 5b^2)$
$3ab(5a - 6b)$	$18a^2b - 10ab^2$	$12a^2b + 16ab^2$
$12a^2b - 18ab^2$	$6b(3a - 2b)$	$15a^2b - 18ab^2$

27. Simplify the fraction shown.
1
2 ▪ 12
3

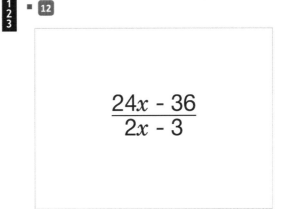

$$\frac{24x - 36}{2x - 3}$$

28. Expand and simplify the following expression and
a fully factorise your answer.
b $3(11a + 2) - 4(7a + 4)$
c

▪ 5(-2 + a) ▪ 5(a - 2)

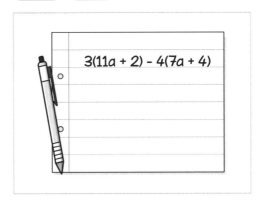

3(11a + 2) - 4(7a + 4)

29. Given that $yb + 5b = b(y + 5)$, what is $y(x + 3) + 5(x$
a $+ 3)$ as a product of two expressions?
b
c ▪ (x + 3)(5 + y) ▪ (5 + y)(x + 3) ▪ (x + 3)(y + 5)
▪ (y + 5)(x + 3)

y(x + 3) + 5(x + 3)

30. Complete the multiplication grid. What expression goes in the striped box?

- -14b + 21ab   - 21ab - 14b

×		
6a	$12a^2 + 30a$	$18a^2 - 12a$
	$14ab + 35b$	

Solve Linear Equations with Unknowns on Both Sides

Competency: Use algebraic methods to solve linear equations in one variable (including all forms that require rearrangement).

Quick Search Ref: 10237

Correct: Correct.   **Wrong:** Incorrect, try again.   **Open:** Thank you.

Level 1:   Understanding - Positive coefficients and positive integer solutions.

✹ **Required:** 7/10   ✹ **Student Navigation:** on   ✹ **Randomised:** off

1. Which is the correct first step for solving the equation $5x + 3 = 2x + 12$?

   ☐
   ☒
   ☐

   ■ Solution A ■ **Solution B** ■ Solution C ■ Solution D

   1/4

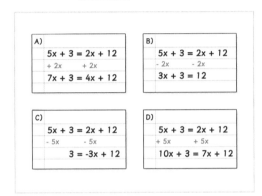

2. Solve the following equation to find the value of a:
   $2a + 4 = a + 7$

   1 2 3 ■ **3**

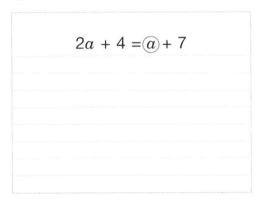

3. If $8x + 2 = 6x + 10$, what is the value of x?

   1 2 3 ■ **4**

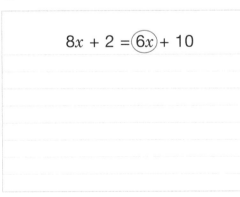

4. Solve the following equation to find the value of x:
   $9x - 4 = 5x + 12$

   1 2 3 ■ **4**

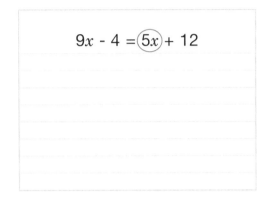

5. Find the value of w if $2w + 4 = 5w + 1$.

   1 2 3 ■ **1**

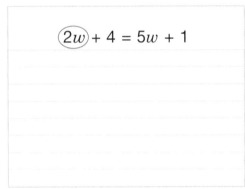

6. If $4t + 7 = 6t - 9$, what is the value of t?

   1 2 3 ■ **8**

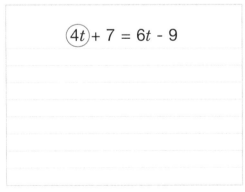

Level 1: *cont.*

7. Solve the following equation to find the value of *m*:
   $9m - 15 = 6m - 3$

   ▪ 4

$$9m - 15 = \boxed{6m} - 3$$

8. Find the value of *x* if $9x + 1 = 2x + 36$.

   ▪ 5

$$9x + 1 = \boxed{2x} + 36$$

9. Find the value of *y*:
   $3y - 4 = y + 8$

   ▪ 6

$$3y - 4 = \boxed{y} + 8$$

10. Solve the following equation to find the value of *c:*
    $c + 19 = 7c - 5$

    ▪ 4

$$\boxed{c} + 19 = 7c - 5$$

Level 2: Fluency - Positive and negative coefficients and positive, negative and fractional answers.

✿ **Required:** 7/10   ✿ **Student Navigation:** on
✿ **Randomised:** off

11. Solve the following equation:
    $5n + 3 = 3n + 1$

    ▪ -1

$$5n + 3 = 3n + 1$$

12. Solve the following equation:
    $12x - 1 = 4x + 5$
    Give your answer as a fraction in its simplest form.

    ▪ 3/4

$$12x - 1 = 4x + 5$$

13. Which is the correct first step for solving the equation $3x - 15 = 20 - 2x$?

1/4

■ Solution A ■ Solution B ■ Solution C ■ Solution D

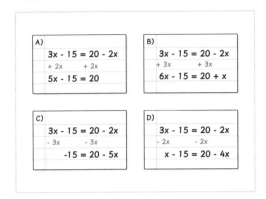

A)
$$3x - 15 = 20 - 2x$$
$$+ 2x \qquad + 2x$$
$$5x - 15 = 20$$

B)
$$3x - 15 = 20 - 2x$$
$$+ 3x \qquad + 3x$$
$$6x - 15 = 20 + x$$

C)
$$3x - 15 = 20 - 2x$$
$$- 3x \qquad - 3x$$
$$-15 = 20 - 5x$$

D)
$$3x - 15 = 20 - 2x$$
$$- 2x \qquad - 2x$$
$$x - 15 = 20 - 4x$$

14. Solve the following equation to find the value of g:
$4g - 3 = 15 - 2g$

 ■ 3

$$4g - 3 = 15 - 2g$$

15. Find the value of k if $2 - 3k = 6k + 47$.

 ■ -5

$$2 - 3k = 6k + 47$$

16. Solve $31 - 5s = 7 - 3s$.

 ■ 12

$$31 - 5s = 7 - 3s$$

17. Solve $10 - 5r = 12 - 9r$.
Give your answer as a fraction in its simplest form.

 ■ 1/2

$$10 - 5r = 12 - 9r$$

18. Solve $3x + 13 = 7x - 5$.
Give your answer as a decimal.

 ■ 4.5

$$3x + 13 = 7x - 5$$

Level 2: *cont.*

19. Solve the following equation to find the value of *y*:

 7*y* + 5 = 27 - 4*y*

■

$$7y + 5 = 27 - 4y$$

20. Solve the following equation and find the value of

 y:

5 - 2*y* = 17 - 5*y*

■ 4

$$5 - 2y = 17 - 5y$$

Level 3: Reasoning - Misconceptions and inverse
         questions.

❋ **Required:** 5/5   ❋ **Student Navigation:** on

❋ **Randomised:** off

21. 2*x* + 7 = 2*x* + 5

a
b   Is it possible to solve the equation? Explain your
c   answer.

$$2x + 7 = 2x + 5$$

22. Kian has correctly solved the equation

a   3*y* + 40 = 10 - 7*y*.
b   Explain what simpler method Kian could have
c   used.

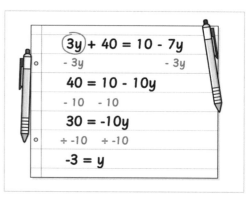

23. Amy has solved an equation using the following
steps:

1. Subtract 4*x* from both sides.

2. Add 5 to both sides.

1/4 3. Divide both sides by 5.

If she gets the solution *x* = 7, what equation did
she start with?

■ 9*x* + 5 = 4*x* + 40   ■ 5*x* + 5 = 4*x* + 12   ■ 9*x* - 5 = 4*x* + 30

■ 5*x* - 12 = 4*x* - 5

24. Paul has made a mistake solving the following

equation:

5*n* - 17 = 25 - 2*n*.

What should the value of *n* be?

■ 6

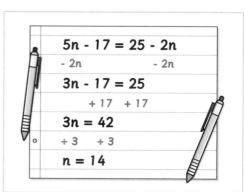

25. Jordan says there are no solutions to the following

a   equation:
b   3(2*x* - 1) + 5 = 2(3*x* + 1)
c   Is Jordan correct? Explain your answer.

$$3(2x - 1) + 5 = 2(3x + 1)$$

Level 4: Problem Solving - Multi-step problems with equations with unknowns on both sides.

✸ **Required:** 5/5   ✸ **Student Navigation:** on
✸ **Randomised:** off

26. Find the size of angle *DEB*.
Do not include units in your answer.

1 2 3 ▪ 115

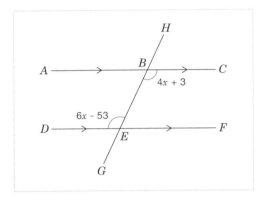

27. Which one of the following equations has the same solution as 3x - 7 = 5x - 3?

▪ 4x + 6 = 7x + 9 ▪ 9x + 4 = 2x - 17 ▪ 10x + 8 = 7x + 2

1/4 ▪ 6x - 8 = 2x + 4

$$3x - 7 = 5x - 3$$

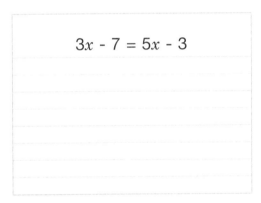

28. Ali thinks of a number. If he doubles his number and adds 9, he gets the same answer as multiplying his number by 5 and subtracting 6. What number is Ali thinking of?

1 2 3

▪ 5

29. Each box in the pyramid is the sum of the expressions in the two boxes below it. Calculate the value of *x*.

1 2 3

▪ 7

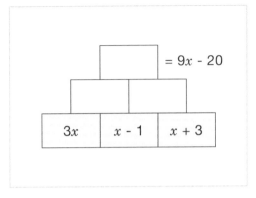

30. What is the side length of the square?

1 2 3 ▪ 42

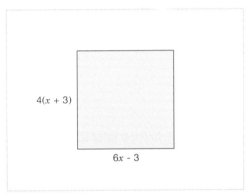

Solve Multi-Step Linear Equations

Competency: Use algebraic methods to solve linear equations in one variable (including all forms that require rearrangement).

Quick Search Ref: **10320**

Correct: Correct. **Wrong:** Incorrect, try again. **Open:** Thank you.

Level 1: Understanding - Positive integer answers and variables with positive coefficients.

✿ **Required:** 7/10 ✿ **Student Navigation:** on ✿ **Randomised:** off

1. Solve the following equation to find the value of
$\begin{smallmatrix}1\\2\\3\end{smallmatrix}$ *m*:
7*m* - 4 = 17

▪ **3**

$$7m - 4 = 17$$
$$+ 4 \qquad + 4$$

2. If *x*/2 + 9 = 12, find the value of *x*.
$\begin{smallmatrix}1\\2\\3\end{smallmatrix}$ ▪ **6**

$$\frac{x}{2} + 9 = 12$$
$$- 9 \qquad - 9$$

3. What is the value of *c* in the equation?
$\begin{smallmatrix}1\\2\\3\end{smallmatrix}$ (*c* + 5)/2 = 12

▪ **19**

$$\frac{c + 5}{2} = 12$$
$$\times 2 \qquad \times 2$$

4. Find the value of *v* if 4(2*v* - 1) = 28.
$\begin{smallmatrix}1\\2\\3\end{smallmatrix}$ ▪ **4**

$$4(2v - 1) = 28$$
$$\div 4 \qquad \div 4$$

5. If 2*x*/3 - 8 = 14, find the value of *x*.
$\begin{smallmatrix}1\\2\\3\end{smallmatrix}$ ▪ **33**

$$\frac{2x}{3} - 8 = 14$$
$$+ 8 \qquad + 8$$

6. Solve the following equation to find the value of *s*:
$\begin{smallmatrix}1\\2\\3\end{smallmatrix}$ 4(2*s* + 3) - 7 = 21

▪ **2**

$$4(2s + 3) - 7 = 21$$
$$+ 7 \qquad + 7$$

Level 1: cont.

7. What is the value of *y* in the equation?

$\frac{1}{2}$ (½y - 3)/5 = 3

■ 36

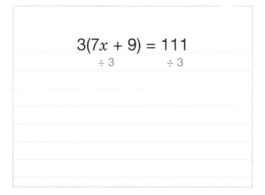

$$\frac{\frac{1}{2}y - 3}{5} = 3$$
$$\times 5 \qquad \times 5$$

8. Find the value of *x* if 3(7x + 9) = 111.

$\frac{1}{2}$ ■ 4

$$3(7x + 9) = 111$$
$$\div 3 \qquad \div 3$$

9. Solve the following equation to find the value of
 w:
2(9w - 7) + 12 = 34

■ 2

$$2(9w - 7) + 12 = 34$$
$$- 12 \qquad -12$$

10. What is the value of *z* in the equation?

$\frac{1}{2}$ 3z/4 - 8 = 1

■ 12

$$\frac{3z}{4} - 8 = 1$$
$$+ 8 \quad + 8$$

Level 2: Fluency - Decimal, fractional and negative
answers and variables with negative
coefficients.

✿ **Required:** 7/10   ✿ **Student Navigation:** on
✿ **Randomised:** off

11. What is the value of *x* if 4x - 3 = 15?
$\frac{1}{2}$ *Give your answer as a decimal.*

■ 4.5

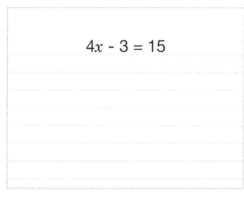

$$4x - 3 = 15$$

12. Solve the following equation to find the value of *b*:
$\frac{1}{2}$ 2(3b + 7) = 8

■ -1

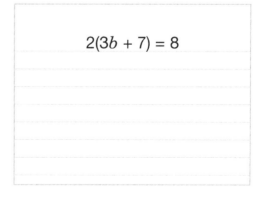

$$2(3b + 7) = 8$$

Level 2: *cont.*

13. What is the value of *n* if 7 - 3*n* = 1?

1
2
3

▪ **2**

$$7 - 3n = 1$$

14. Find the value of *r* in the following equation:
a
b
c
3*r*/2 - 14 = 6
Give your answer as a mixed number fraction.

▪ **13 1/3**

$$\frac{3r}{2} - 14 = 6$$

15. Solve the following equation to find the value of *x*:
a
b
c
2(7*x* + 5) = 17
Give your answer as a fraction in its simplest form.

▪ **1/2**

$$2(7x + 5) = 17$$

16. Find the value of *m* in the following equation:
1
2
3
3(9 - 2*m*) + 11 = 50

▪ **-2**

$$3(9 - 2m) + 11 = 50$$

17. Solve the equation to find the value of *a*.
1
2
3
(4*a* + 11)/3 = 15
Give your answer as a decimal.

▪ **8.5**

$$\frac{4a + 11}{3} = 15$$

18. Find the value of *x* in the following equation:
a
b
c
2(6*x* - 5) + 17 = 49
Give your answer as a mixed number fraction.

▪ **3 1/2**

$$2(6x - 5) + 17 = 49$$

19. Solve the following equation to find the value of *t*.

1
2
3

$(5t + 12)/3 = 18$

Give your answer as a decimal.

▪ 8.4

$$\frac{5t + 12}{3} = 18$$

20. Find the value of *n* in the following equation.

1
2
3

$(2n + 11)/3 = 1$

▪ -4

$$\frac{2n + 11}{3} = 1$$

Level 3: Reasoning - Inverse questions and misconceptions solving multi-step equations.

✿ Required: 5/5   ✿ Student Navigation: on
✿ Randomised: off

21. Oliver has made a mistake solving the equation. Explain where Oliver made his mistake.

a
b
c

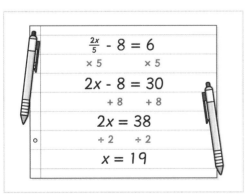

$$\frac{2x}{5} - 8 = 6$$
$$\times 5 \qquad \times 5$$
$$2x - 8 = 30$$
$$+ 8 \qquad + 8$$
$$2x = 38$$
$$\div 2 \qquad \div 2$$
$$x = 19$$

22. Owen thinks of a number, multiplies it by 7, subtracts 9 and then doubles it.
If he gets an answer of 150, what number did Owen first think of?

1
2
3

▪ 12

23. Aisha says the solution to $x^2 + 9 = 25$ is 4.
What other solution is there to this equation?

a
b
c

▪ -4

24. Tamsin is solving the equation $5(2w - 7) = 17$ and begins by expanding the bracket.
Eden thinks Tamsin has made a mistake and her first step should have been to divide both sides by 5.
Is Eden correct? Explain your answer.

a
b
c

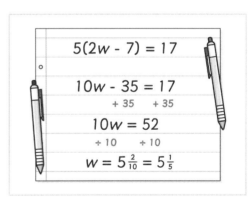

$$5(2w - 7) = 17$$
$$10w - 35 = 17$$
$$+ 35 \qquad + 35$$
$$10w = 52$$
$$\div 10 \qquad \div 10$$
$$w = 5\frac{2}{10} = 5\frac{1}{5}$$

25. Select three possible values of *n* so that the equation has an integer solution for *x*.

▪ 0 ▪ 1 ▪ 2 ▪ 5 ▪ 116 ▪ 231 ▪ 378

3/7

$$\frac{2x}{3} + 5 = n$$

Level 4: Problem Solving - Multi-step problems solving equations.

Required: 5/5 **Student Navigation:** on

Randomised: off

26. Arrange the equations in ascending order according to the value of x, smallest first.

- 3(5x + 12) + 4 = 10
- 4(6x - 1) = 8
- 2(8 - 3x) = 10
- 7(3x + 2) = 63

$$4(6x - 1) = 8$$

$$7(3x + 2) = 63$$

$$2(8 - 3x) = 10$$

$$3(5x + 12) + 4 = 10$$

27. The area of the shape is 118 cm².
Calculate the perimeter of the shape in centimetres (cm).
Include units in your answer.

- 58 cm
- 58 centimetres

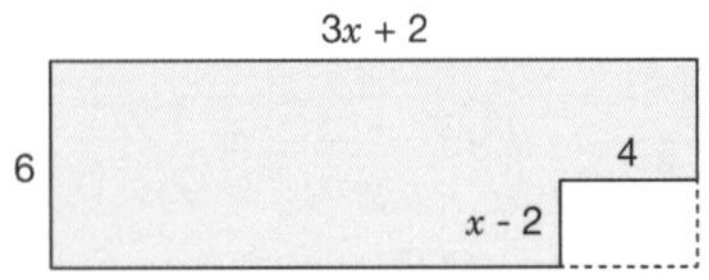

28. What is the size in degrees of the largest angle in the triangle?
Don't include units in your answer.

- 95

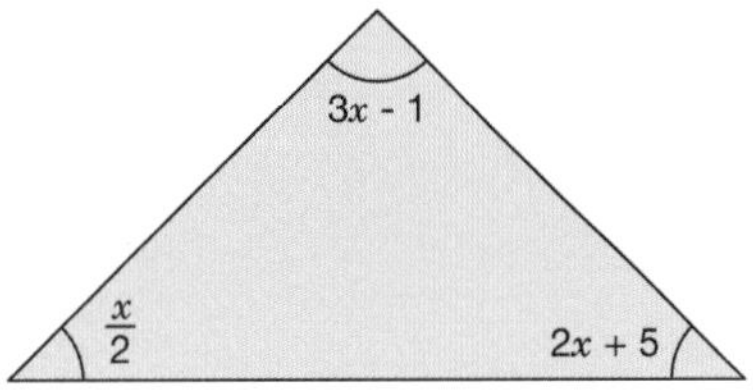

29. Sam, Seb, Sara and Sonny have 181 marbles between them.
Sam has s marbles, Sara has 7 more than Sam, Seb has half as many as Sara and Sonny has 10 less than Seb.
How many marbles does Seb have?

- 33

30. The perimeter of the rectangle is 32 cm.
Find the value of x.
Don't include units in your answer.

- 6

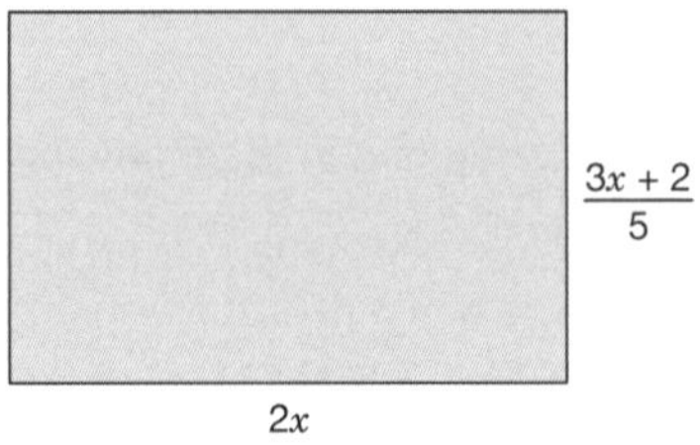

Solve linear inequalities in one variable

Competency: Understand and use the concepts and vocabulary of expressions, equations, inequalities, terms and factors.

Quick Search Ref: 10109

Correct: Correct.   Wrong: Incorrect, try again.   Open: Thank you.

Level 1: Understanding - Solving simple inequalities.

✿ **Required:** 7/10   ✿ **Student Navigation:** on   ✿ **Randomised:** off

---

1. Arrange the following symbols in the same order as the definitions shown.

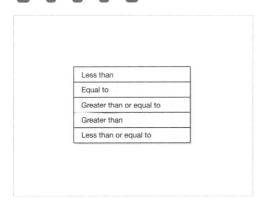

2. If $3x < 48$, which inequality represents the value of x?

- $x < 45$ - $x < 48$ - **x < 16**

1/3

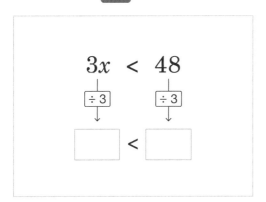

3. Given that $k + 7 \geq 18$, which inequality represents the value of k?

- $k \geq 18$ - $k \geq 25$ - **k ≥ 11**

1/3

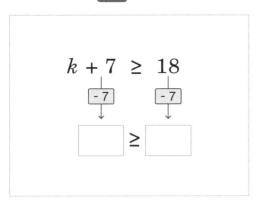

4. If $6d < 24$, what inequality represents d?
Use the < and > keys to type the less than and greater than symbols.

- **d < 4**

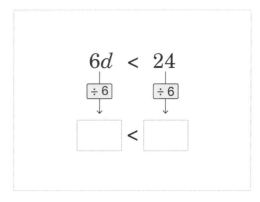

5. Given that $r - 9 > 14$, what inequality represents r?
Use the < and > keys to type the less than and greater than symbols.

- **r > 23**

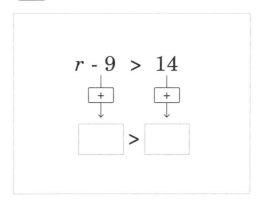

6. If $4a - 8 < 36$, what inequality represents a?
Use the < and > keys to type the less than and greater than symbols.

- **a < 11**

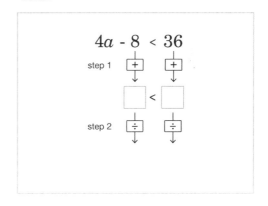

Level 1: *cont.*

7. Solve the inequality

a
b $3(n + 5) < 18$.
c *Use the < and > keys to type the less than and greater than symbols.*

▪ n < 1

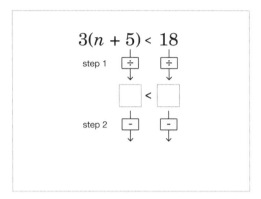

8. Given that $f + 18 > 31$, what inequality represents

a f?
b *Use the < and > keys to type the less than and*
c *greater than symbols.*

▪ f > 13

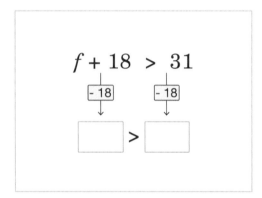

9. If $7c + 8 < 57$, what inequality represents c?

a *Use the < and > keys to type the less than and*
b *greater than symbols.*
c

▪ c < 7

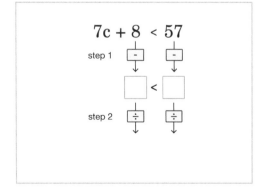

10. Solve the inequality

a $2(g - 7) > 32$.
b *Use the < and > keys to type the less than and*
c *greater than symbols.*

▪ g > 23

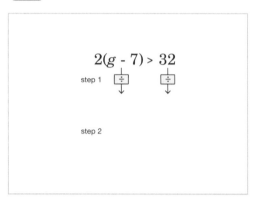

Level 2: Fluency - Solving inequalities involving fractions and compound inequalities.

✺ **Required:** 7/10   ✺ **Student Navigation:** on
✺ **Randomised:** off

11. Solve $y/3 - 9 < 4$.

a ▪ y < 39
b
c

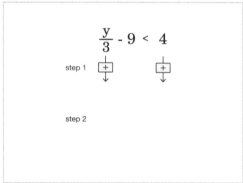

12. Solve $(t + 12)/5 > 8$.

a ▪ t > 28
b
c

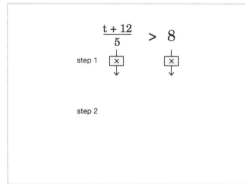

13. Select the integers that satisfy the compound inequality

$3 < x + 1 \le 6$.

3/6   ▪ 7 ▪ 5 ▪ 6 ▪ 3 ▪ 2 ▪ 4

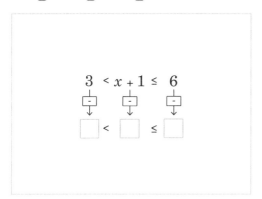

14. Select the integers that satisfy the compound inequality

$6 \le 2e < 12$.

3/6   ▪ 5 ▪ 12 ▪ 4 ▪ 6 ▪ 2 ▪ 3

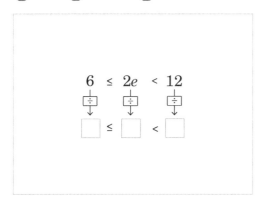

15. If $1 < h - 7 < 8$, what compound inequality represents h?

a b c   ▪ 8 < h < 15

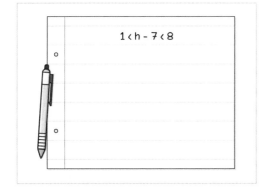

16. Solve the inequality

a b c $25 < 2z + 7 < 45$.

▪ 9 < z < 19

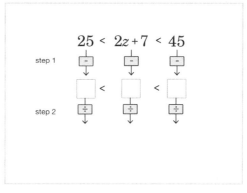

17. Solve $-3 < w/2 - 7 < 5$.

a b c   ▪ 8 < w < 24

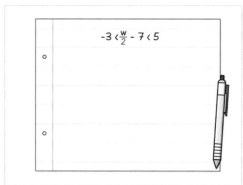

18. If $-6 < v - 4 < 12$, what compound inequality represents v?

a b c   ▪ -2 < v < 16

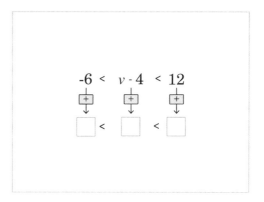

Level 2: *cont.*

19. Solve the inequality

$1 < p/3 + 5 < 7$.

- $-12 < p < 6$

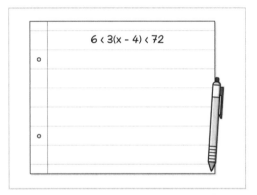

20. Solve $6 < 3(x - 4) < 72$.

- $6 < x < 28$

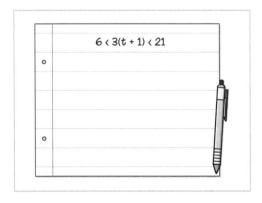

Level 3: Reasoning - Applying linear inequalities.

✱ **Required:** 5/5   ✱ **Student Navigation:** on
✱ **Randomised:** off

21. Meg says there are four solutions to the inequality
$6 < 3(t + 1) < 21$.
Is Meg correct? Explain your answer.

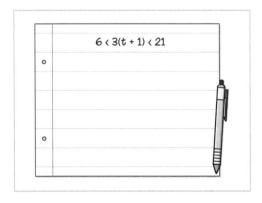

22. Which number line represents the solution to the
following inequality?
$12 < 4(x + 1) < 24$

1/4   ■ (i)   ■ (ii)   ■ (iii)   ■ (iv)

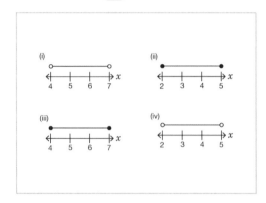

23. How many integers satisfy the following
inequality?
$4 < 2b - 1 \leq 10$

- 3

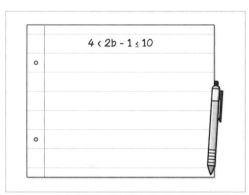

24. Joseph is solving $-x < -2$ and caculates that $x < 2$. Is
Joseph correct? Explain your answer.

25. Georgia has made one mistake in her solving
a inequalities homework. What is the correct
b answer to the question she got wrong?
c

- -5 < x < 2

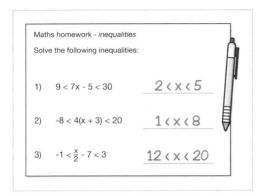

Level 4: Problem Solving - Multi-step problems
involving linear inequalities.

✴ **Required:** 5/5   ✴ **Student Navigation:** on
✴ **Randomised:** off

26. Which two inequalities have the same solution
☐ set?
☒
☐
2/5
- 0 < 4(x + 5) < 32 ■ 5 < x/3 + 4 < 7 ■ 4 < 5x - 1 < 39
- -1 < 2x - 7 < 11 ■ -2 < x/2 - 4 < 3

27. The grid contains four pairs of equivalent
a compound inequalities. What is the solution to the
b odd one out?
c

- 3 < x < 6

$3 < x < 11$	$5 < \frac{x}{2} + 6 < 9$	$4 < x < 7$
$-2 < x < 6$	$3 < \frac{x+7}{2} < 8$	$-1 < x < 9$
$8 < 3x - 1 < 17$	$-4 < 2(x - 5) < 12$	$5 < 2x - 3 < 11$

28. Write an inequality for the set of values that
a satisfy both inequalities below.
b 15 < 3(m + 4) < 48
c 6 < m/2 + 7 < 10

- -2 < m < 12

29. A crate of apples weighs more than 9 kilograms
a (kg) but less than 12 kilograms.
b If the crate contains 12 identical bags of apples
c and weighs 480 grams (g) when empty, write an
inequality for the weight, *a*, of a single bag of
apples in grams.

- 710 g < a < 960 g ■ 710 < a < 960
- 710 grams < a < 960 grams

30. Alana thinks of a number, multiplies it by 4 and
1 adds 11. Her answer is greater than 3 but less than
2 25. How many integers could Alana have started
3 with?

- 5

Use a number line to represent an inequality and to find integer solutions

Competency: Understand and use the concepts and vocabulary of expressions, equations, inequalities, terms and factors.

Quick Search Ref: 10262

Correct: Correct.   **Wrong:** Incorrect, try again.   **Open:** Thank you.

Level 1:   Understanding - Number lines and single inequalities.

✹ **Required:** 7/10   ✹ **Student Navigation:** on   ✹ **Randomised:** off

1. Arrange the following symbols in the same order as the definitions shown.

   ▪ < ▪ ≥ ▪ > ▪ ≤

   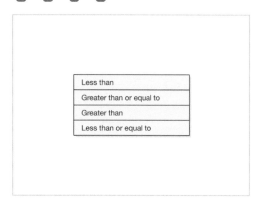

2. Arrange the inequalities so they match the number lines shown.

   ▪ x > -2 ▪ x ≤ -2 ▪ x < -2 ▪ x ≥ -2

   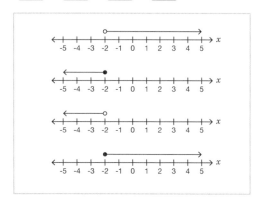

3. Which inequality is represented by the number line shown?

   ▪ x > 3 ▪ x ≤ 3 ▪ x < 3 ▪ x ≥ 3

   1/4

   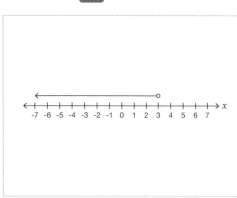

4. Which number line represents the following inequality?

   x ≥ -2

   1/4   ▪ (i) ▪ (ii) ▪ (iii) ▪ (iv)

   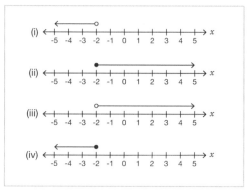

5. Select two values that satisfy the inequality n > -1.

   ▪ 0 ▪ -1 ▪ -3 ▪ 3 ▪ -2

   2/5

   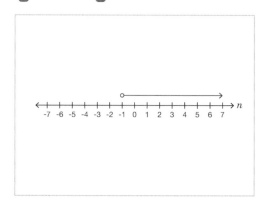

6. How many possible positive integer values are there for x?

   x < 3

   ▪ 2

   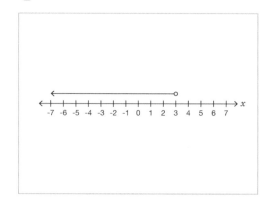

Level 1: cont.

7. What inequality is represented by the number line shown?

a
b
c

Use the < and > keys to write the less than and greater than symbols.

▪ x > -3 ▪ -3 < x

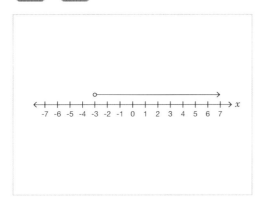

8. What inequality is represented by the number line shown?

a
b
c

Use the < and > keys to type the less than and greater than symbols.

▪ 2 > x ▪ x < 2

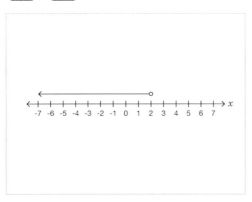

9. If $t > -2$, what is the smallest possible integer t can be?

a
b
c

▪ -1

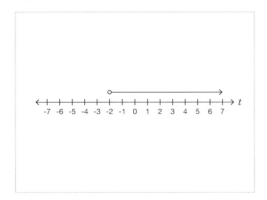

10. Select two values that satisfy the inequality $x \geq 2$.

▪ -1 ▪ 2 ▪ -3 ▪ 4 ▪ 1

2/5

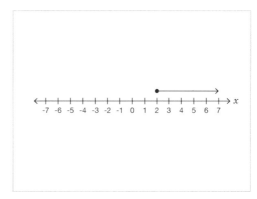

Level 2: Fluency - Number lines and compound inequalities.

✿ **Required:** 7/10   ✿ **Student Navigation:** on
✿ **Randomised:** off

11. Select all the integers that satisfy both the inequalities below:
$x < 4$ and $x \geq 1$

3/7   ▪ 0 ▪ 1 ▪ 2 ▪ 3 ▪ 4 ▪ 5 ▪ 6

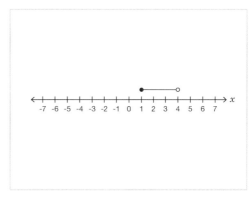

12. Which number line represents the compound inequality
$x \geq -2$ and $x < 4$?

1/4   ▪ (i) ▪ (ii) ▪ (iii) ▪ (iv)

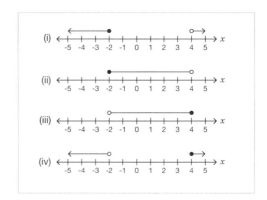

13. Select the integers that satisfy the compound inequality
$a < -1$ or $a > 4$.

3/6 ■ -2 ■ 0 ■ 4 ■ 2 ■ 6 ■ -4

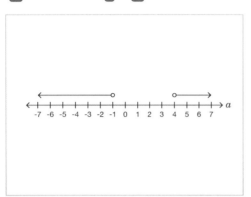

14. Which compound inequality is represented by the number line shown?

1/4 ■ $-2 \le x < 1$ ■ $-2 < x \ge 1$ ■ $-2 \le x > 1$ ■ $-2 < x \le 1$

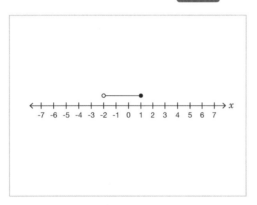

15. What compound inequality is represented by the number line shown?
Use the < and > keys to type the less than and greater than symbols.

■ $-1 < d < 3$ ■ $3 > d > -1$

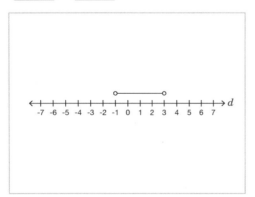

16. What is the largest possible **even number** which satisfies the inequality
$-3 \le x < 6$?

■ 4

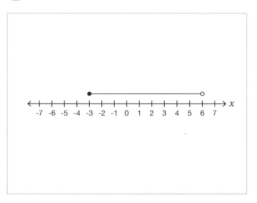

17. What compound inequality is represented by the number line shown?
Use the < and > keys to type the less than and greater than symbols.

■ $-3 < n < 2$ ■ $2 > n > -3$

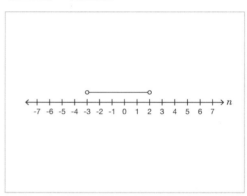

18. How many integers are there that satisfy the following compound inequality:
$-2 < z < 4$?

■ 5

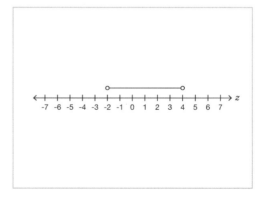

Level 2: cont.

19. Select all integer values that satisfy the compound inequality
-2 < x ≤ 1.

3/6 ▪ 1 ▪ -2 ▪ 0 ▪ -3 ▪ 2 ▪ -1

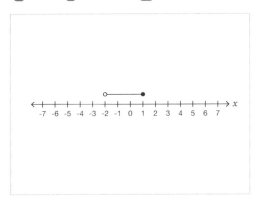

20. What is the smallest possible **odd** integer which satisfies the compound inequality
-3 < x < 2?

▪ -1

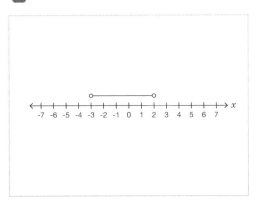

Level 3: Reasoning - Using inequalities to solve problems.

✿ **Required:** 5/5   ✿ **Student Navigation:** on
✿ **Randomised:** off

21. Which compound inequality has the most integer solutions?

▪ -2 ≤ x < 2 ▪ x < -1 or x ≥ 3 ▪ x ≥ 1 and x ≤ 5

1/3

22. In the inequalities r > 7 and s ≥ 9, r and s are both integers.
What is the smallest possible value of r × s?

▪ 72

23. Koby made one mistake in his inequalities homework. What is the correct answer to the question he got wrong?
Use the < and > keys to type the less than and greater than symbols.

▪ -1 < b < 3

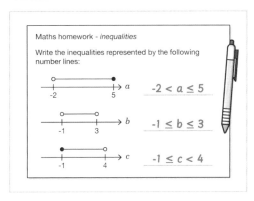

24. Which inequality only has integer solutions of -3, -2, -1, 0 and 1?

▪ -3 < x ≤ 1 ▪ x > -4 and x < 1 ▪ -4 ≤ x ≤ 1
1/5 ▪ x < 2 and x ≥ -3 ▪ 2 ≥ x ≥ -3

25. Lois says there are 7 odd integer solutions that satisfy the compound inequality
-3 < n < 11.
Is Lois correct? Explain your answer.

Level 4: Problem Solving - Multi-step problems using number lines and inequalities.

✿ **Required:** 5/5   ✿ **Student Navigation:** on
✿ **Randomised:** off

26. Olivia is plotting points such that the x and y co-ordinates are all integers and
-2 ≤ x < 5 and -1 < y < 6.
How many different points could Olivia plot?

▪ 42

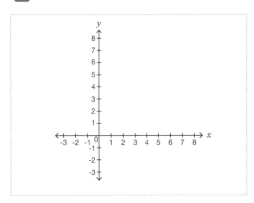

27. n is a square number.
20 < n² < 100
What is the value of n?

▪ 9

Level 4: cont.

28. u and v are integers and
$1 < u \leq 24$ and $3 < v \leq 12$.
What is the greatest possible value of $u \div v$?

- 6

29. How many integers are there that satisfy all three
compound inequalities?
$x < -5$ or $x \geq 3$
$-10 \leq y < 7$
$z > -8$ and $z < 10$

- 6

30. The length of a rectangle is l and the width is w.
$3 < l \leq 8$ and $2 \leq w < 7$
If l and w are integers, how many different areas
could the rectangle have?

- 19

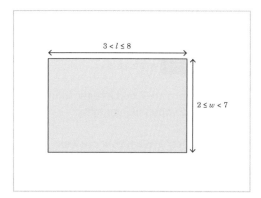

$3 < l \leq 8$

$2 \leq w < 7$

Mathematics

Y8

Measurement

Metric Units

Area and Perimeter

Volume and Capacity

Compound Units

Compare and Convert Metric Units of Area

Competency: Change freely between related standard units (for example time, length, area, volume/capacity, mass).

Quick Search Ref: 10365

Correct: Correct.   **Wrong:** Incorrect, try again.   **Open:** Thank you.

Level 1: Understanding - How to convert between metric units of area.

✿ **Required:** 7/7   ✿ **Student Navigation:** on   ✿ **Randomised:** off

---

1. Sort the options in the order they would appear in the table.

↑
↓

▪ ÷ 10 ▪ ÷100 ▪ ÷ 1,000 ▪ × 1,000 ▪ × 100
▪ × 10

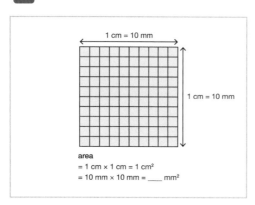

2. How many square millimetres (mm²) are there in 1 square centimetre (cm²)?
Don't include the units in your answer.

1
2
3

▪ 100

3. How do you convert from mm² to cm²?

☐
☒
☐

▪ × 10 ▪ ÷ 1,000 ▪ × 100 ▪ ÷ 10 ▪ ÷ 100

1/5

mm to cm : ÷ 10
mm² to cm² : ÷ 10?

4. How many square centimetres (cm²) are there in one square metre (1 m²)?
Don't include the units in your answer.

a
b
c

▪ 10,000 ▪ 10000

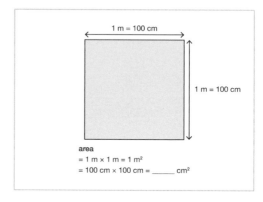

5. How do you convert from cm² to m²?

☐
☒
☐

▪ × 100 ▪ ÷ 1,000 ▪ × 10,000 ▪ ÷ 100 ▪ ÷ 10,000

1/5

cm to m : ÷ 100
cm² to m² : ÷ 100?

6. How many square metres (m²) are there in one square kilometre (km²)?
Don't include the units in your answer.

a
b
c

▪ 1,000,000 ▪ 1000000

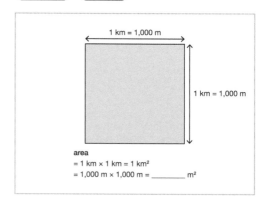

Level 1: *cont.*

7. How do you convert from m² to km²?

1/5

- × 1,000
- ÷ 10,000
- ÷ 1,000,000
- × 10,000
- ÷ 1,000

> m to km   : ÷ 1,000
> m² to km² : ÷ 1,000?

Level 2: Fluency - Converting between metric units of area.

✸ **Required:** 7/10   ✸ **Student Navigation:** on
✸ **Randomised:** off

8. 60,000 cm² = __ m².

- 6

> cm to m   : ÷ 100
> cm² to m² : ____

9. What is 120,300 m² in km²?
Don't include the units in your answer.

- 0.1203

> m to km   : ÷ 1,000
> m² to km² : _____

10. 7 cm² = ___ mm².

- 700

> cm to mm   : × 10
> cm² to mm² : ____

11. Calculate 1.26 km² in m².
Don't include the units in your answer.

- 1260000
- 1,260,000

> km to m   : × 1,000
> km² to m² : _____

12. What is the area of the rectangle in m²?
Don't include the units in your answer.

- 0.035

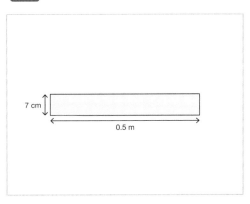

7 cm
0.5 m

13. Joe's bedroom measures 3.5 m² by 6 m². What is the area of his bedroom in cm²?
Don't include the units in your answer.

- 210,000
- 210000

14. A piece of paper measures 21 cm by 29.7 cm. What is the area in mm²?
Don't include the units in your answer.

- 62,370
- 62370

15. 27 km² = ___ m².

a
b
c
■ 27,000,000 ■ 27000000

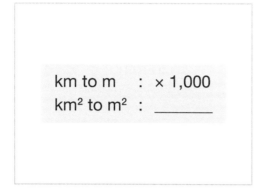

km to m   : × 1,000
km² to m² : _____

16. What is 40,881 mm² in cm²?
Don't include the units in your answer.

a
b
c
■ 408.81

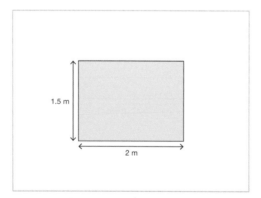

mm to cm : ÷ 10
mm² to cm² : ____

17. What is the area of the rectangle in cm²?
Don't include the units in your answer.

a
b
c
■ 30000 ■ 30,000

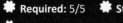

1.5 m

2 m

✱ **Required:** 5/5   ✱ **Student Navigation:** on
✱ **Randomised:** off

18. Toby says, "1 m = 100 cm, so 1 m² = 100 cm²".
Is Toby correct? Explain your answer

a
b
c

19. What is the surface area of the cube in m²?
Don't include the units in your answer.

1
2
3
■ 0.0486

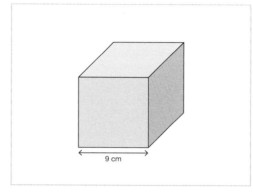

9 cm

20. Select the symbol that makes the following
statement true:
6,400 mm² ___ 0.064 m²

1/3 ■ < ■ = ■ >

21. What is 1 m² in mm²?
Don't include the units in your answer.

a
b
c
■ 1000000 ■ 1,000,000

22. Calculate the area of the triangle in mm².
Don't include the units in your answer.

1
2
3
■ 336

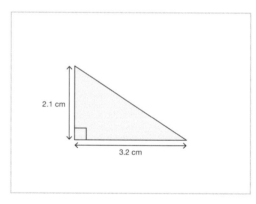

2.1 cm

3.2 cm

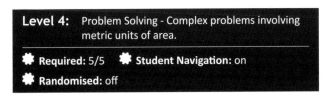

Level 4: Problem Solving - Complex problems involving metric units of area.

✱ **Required:** 5/5   ✱ **Student Navigation:** on
✱ **Randomised:** off

23. If a square has an area of 1,600 mm², what is the length of one side in cm?
Include the units cm in your answer.

a
b
c

▪ 4 cm ▪ 4.0 cm ▪ 4.0 centimetres ▪ 4 centimetres

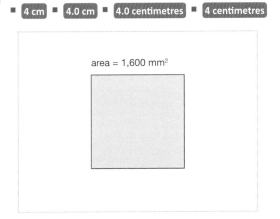

area = 1,600 mm²

24. The scale of a map is 1 cm : 200 m. The park has dimensions of 350 m by 280 m. What is the area of the park on the map in cm²?
Don't include the units in your answer.

a
b
c

▪ 2.45

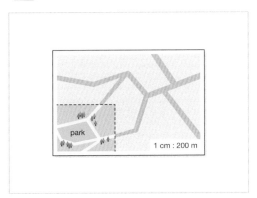

1 cm : 200 m

25. Jan wants to use some lawn feed on the grass in her garden. If a bag of feed covers 80,000 cm², how many bags will she need?

a
b
c

▪ 10

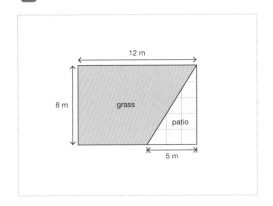

26. Calculate the **volume** of the cuboid in cm³.
Don't include the units in your answer.

a
b
c

▪ 2.88

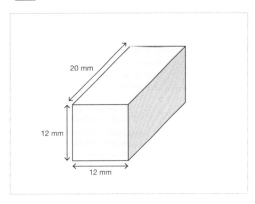

20 mm
12 mm
12 mm

27. A piece of land measures 1.1 km by 1.2 km. If on average a farmer's field has dimensions of 120 m by 110 m, how many fields could the land accommodate?

a
b
c

▪ 100

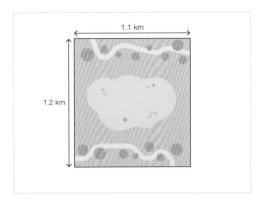

1.1 km
1.2 km

Compare and Convert Metric Units of Length, Mass and Capacity

Competency: Change freely between related standard units (for example time, length, area, volume/capacity, mass).

Quick Search Ref: 10366

Correct: Correct.   Wrong: Incorrect, try again.   Open: Thank you.

Level 1: Understanding - converting between metric units.

✱ **Required:** 7/10   ✱ **Student Navigation:** on   ✱ **Randomised:** off

1. Select the **three** metric units of measure.

■ metre ■ foot ■ kilogram ■ litre ■ pound

3/5

2. Arrange these metric units of length in ascending order according to size (smallest first).

■ millimetre ■ cenitmetre ■ metre ■ kilometre

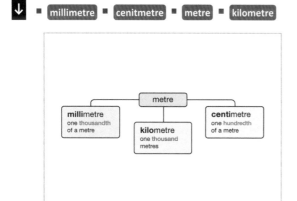

3. When converting from litres to millilitres you:

■ ÷ 1,000 ■ × 1,000 ■ ÷ 100 ■ × 10 ■ × 100

1/5

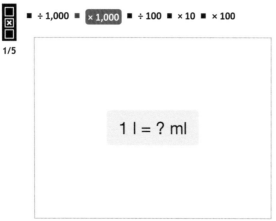

1 l = ? ml

4. When converting from centimetres to millimetres you:

■ ÷ 10 ■ ÷ 100 ■ ÷ 1,000 ■ × 10 ■ × 100

1/5

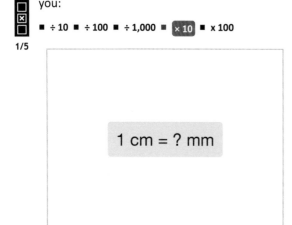

1 cm = ? mm

5. When converting from grams to kilograms you:

■ ÷ 1,000 ■ × 1,000 ■ ÷ 100 ■ × 10 ■ ÷ 10

1/5

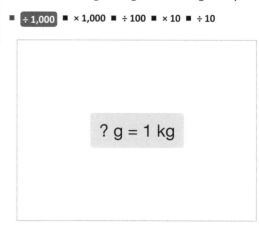

? g = 1 kg

6. Arrange these metric units of capactiy in descending order according to size (largest first).

■ Litre ■ Decilitre ■ Centilitre ■ Millilitre

prefix	value
mega -	× 1,000,000
kilo -	× 1,000
hecto -	× 100
deca -	× 10
base unit	
deci -	÷ 10
centi -	÷ 100
milli -	÷ 1,000
micro -	÷ 10,000,000
nano -	÷ 1,000,000,000

7. When converting from metres to centimetres you:

■ × 1,000 ■ × 100 ■ ÷ 1,000 ■ × 10 ■ ÷ 100

1/5

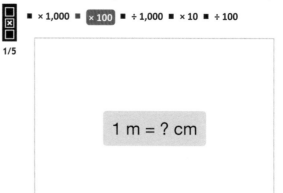

1 m = ? cm

Level 1: *cont.*

8. When converting from kilometres to metres you:

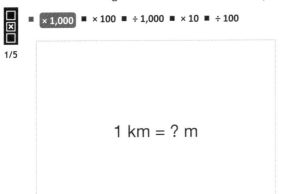

■ ×1,000 ■ ×100 ■ ÷1,000 ■ ×10 ■ ÷100

1/5

$$1 \text{ km} = ? \text{ m}$$

9. When converting from millilitres to litres you:

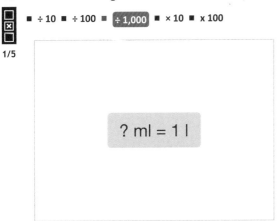

■ ÷10 ■ ÷100 ■ ÷1,000 ■ ×10 ■ x100

1/5

$$? \text{ ml} = 1 \text{ l}$$

10. Arrange these metric units of mass in ascending order according to size (smallest first).

■ Nanogram ■ Milligram ■ Gram ■ Kilogram

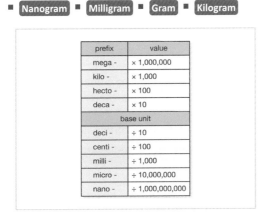

prefix	value
mega -	× 1,000,000
kilo -	× 1,000
hecto -	× 100
deca -	× 10
base unit	
deci -	÷ 10
centi -	÷ 100
milli -	÷ 1,000
micro -	÷ 10,000,000
nano -	÷ 1,000,000,000

Level 2: Fluency - Converting between metric units in context.

✱ Required: 7/10   ✱ Student Navigation: on
✱ Randomised: off

11. What is the capacity of the carton in litres?
Include the units l (litres) in your answer.

■ 0.430 litres ■ 0.43 litres ■ 0.430 l ■ 0.43 l

430 ml

12. Arrange the lengths in ascending order according to size (smallest first).

■ 891 mm ■ 12 m ■ 2,842 cm ■ 143 km

13. 8.45 kg + 5,349 g = _____ kg.

■ 13.799

$$8.45 \text{ kg} + 5,349 \text{ g} = \text{_____ kg}$$

14. 200 cl (centilitres) of water is needed for an experiment. If Nicky repeats the experiment 12 times how many litres of water does he need in total?
Include the unit l (litres) in your answer.

■ 24 litres ■ 24 l

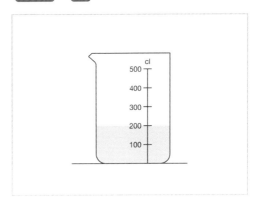

cl
500
400
300
200
100

Level 2: *cont.*

15. A football pitch is 0.12 km in length and 55 m in
width. In a training session Lucy runs around the
perimeter of the pitch seven times. How many
kilometres does Lucy run?
Include the units km (kilometres) in your answer.

- 2.45 km - 2.450 km - 2.45 kilometres
- 2.450 kilometres

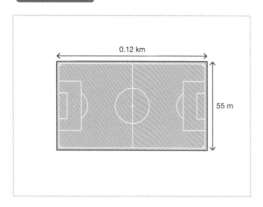

16. A recycling plant collects 63.4 kg of plastic in 80
containers. What is the average mass of plastic in
each container?
Include the units g (grams) in your answer.

- 792.5 g - 792.5 grams

17. At the school sports day Evan drinks two half-
litre bottles of water and Jenny drinks five 250 ml
cups of water. How much more water does Jenny
drink than Evan?
Include the units l (litres) in your answer.

- 0.25 litres - 0.250 litres - 0.25 l - 0.250 l

18. 12.3 l - 6,322 ml = _____ l.

- 5.978

19. Arrange the lengths in descending order according
to size (largest first).

- 7.2 kg - 0.42 kg - 417 g - 12,459 mg

20. What is the mass of the bag of potatoes in grams?
Include the units g (grams) in your answer.

- 2420 grams - 2,420 grams - 2420 g - 2,420 g

Level 3: Reasoning - Understanding practical uses of
metric units.

❉ **Required:** 5/5   ❉ **Student Navigation:** on
❉ **Randomised:** off

21. What is a practical use for centilitres?

22. What symbol makes the following statement true?
19,482 mg ___ 0.02 kg.

- < - = - >

1/3

23. Maisie says, "A 75 cl bottle is equal to 7.5 litres". Is
she correct? Explain your answer.

24. A 22.7 m long path is made up of 40 identical
paving stones. If 3 paving stones are removed how
much has the path been shortened by in
centimetres?
Include the units cm (centimetres) in your answer.

- 170.25 centimetres - 170.25 cm

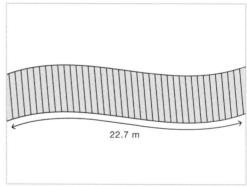

Level 3: *cont.*

25. What metric units would you use to estimate the journey from Madrid to Barcelona? Explain why.

a b c

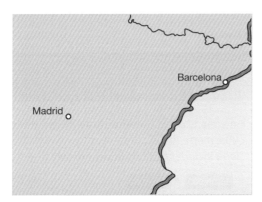

Level 4: Problem Solving - Converting between metric units in worded problems.

✱ **Required:** 5/5   ✱ **Student Navigation:** on
✱ **Randomised:** off

26. Carpet tiles are 41 cm by 30 cm. Ryder uses 48 tiles to cover his floor. What is the area of his floor in m²?

a b c

Don't include the units in your answer.

▪ 5.904

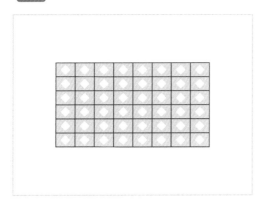

27. 2.24 litres of a mixed fruit juice contains orange, raspberry and apple juice in the ratio 1:2:4. How much raspberry juice is needed to make the mixed fruit juice?

a b c

Include the units ml (millilitres) in your answer.

▪ 640 millilites ▪ 640 ml

28. In an addition pyramid each box is the sum of the two boxes below. What measurement goes in the highlighted box?

a b c

Give your answer rounded to 1 decimal place and include the units kg (kilograms) in your answer.

▪ 5.2 kg ▪ 5.2 kilograms

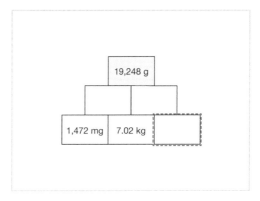

29. Five bags of sugar have the same mass as eight bags of flour. What is the mass of one bag of flour in grams?

a b c

Include the units g (grams) in your answer.

▪ 525 grams ▪ 525 g

30. Spencer is making the 324.12 km journey from Edinburgh to York. He travels two-thirds of the total distance by train and a quarter of the total distance by car. He then travels 1/32 of the remaining journey on foot. How far does Spencer walk in metres?

a b c

Give your answer to the nearest whole number and include the units m (metres) in your answer.

▪ 844 m ▪ 844 metres

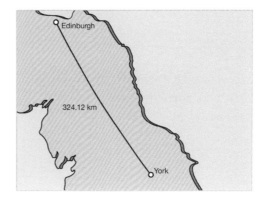

Compare and Convert Metric Units of Volume

Competency: Change freely between related standard units [for example time, length, area, volume/capacity, mass].

Quick Search Ref: 10389

Correct: Correct.   **Wrong:** Incorrect, try again.   **Open:** Thank you.

Level 1: Understanding - How to convert between metric units of volume.

✿ **Required:** 7/7   ✿ **Student Navigation:** on   ✿ **Randomised:** off

---

1. Calculate the volume of the cube in cm³.
Don't include the units in your answer.

▪ 125

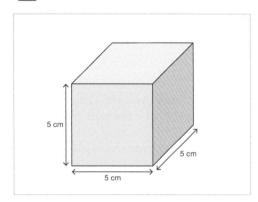

2. How many cubic millimetres (mm³) are there in 1 cubic centimetre (cm³)?
Don't include the units in your answer.

▪ 1,000 ▪ 1000

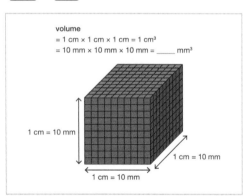

3. How do you convert from mm³ to cm³?

▪ × 1,000 ▪ ÷ 10 ▪ ÷ 1,000 ▪ × 10 ▪ ÷ 100

1/5

4. How many cubic centimetres (cm³) are there in 1 cubic metre (m³)?
Don't include the units in your answer.

▪ 1,000,000 ▪ 1000000

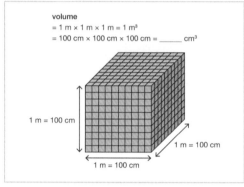

5. How do you convert from cm³ to m³?

▪ ÷ 1,000,000 ▪ ÷ 1,000 ▪ × 1,000 ▪ ÷ 10,000 ▪ × 100

1/5

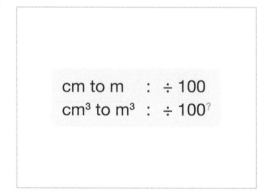

6. 1,000 millilitres = __ litre.

▪ 1

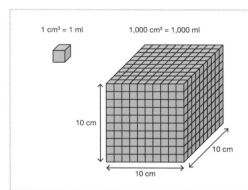

Level 1: *cont.*

7. 1 litre = _____ cm³.

a
b
c
▪ 1,000 ▪ 1000

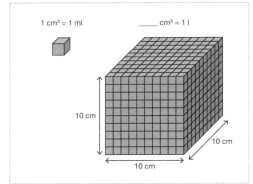

1 cm³ = 1 ml   _____ cm³ = 1 l

10 cm
10 cm
10 cm

Level 2: Fluency - Converting between metric units of volume.

✱ **Required:** 7/10   ✱ **Student Navigation:** on
✱ **Randomised:** off

8. 16,000,000 cm³ = _____ m³.

a
b
c
▪ 16

cm to m   : ÷ 100
cm³ to m³ : _____

9. 18.7 cm³ = ___ mm³.

a
b
c
▪ 18,700 ▪ 18700

cm to mm   : × 10
cm³ to mm³ : _____

10. What is 40,881 mm³ in cm³?

a
b
c
Don't include the units in your answer.

▪ 40.881

mm to cm   : ÷ 10
mm³ to cm³ : ____

11. Convert 0.57 m³ to cm³.

a
b
c
Don't include the units in your answer.

▪ 570000 ▪ 570,000

m to cm   : × 100
m³ to cm³ : _____

12. What is the volume of the cuboid in cm³.

1
2
3
Don't include the units in your answer.

▪ 4

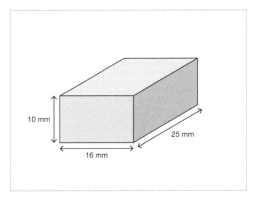

10 mm
25 mm
16 mm

Level 2: *cont.*

13. A bottle of water holds 1.5 litres. What is its
a volume in cm³?
b *Don't include the units in your answer.*
c

   ▪ 1,500 ▪ 1500

14. A container measures 20 cm x 60 cm x 80 cm.
a What is the maximum volume of liquid it will hold
b in m³?
c *Don't include the units in your answer.*

   ▪ 0.096

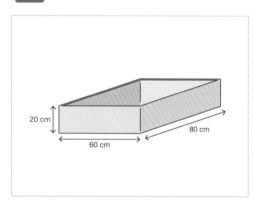

15. Jake has 500 cm³ of water. What is this in litres?
a *Give your answer as a decimal and include the*
b *units l (litres) in your answer.*
c

   ▪ 0.5 l ▪ 0.5 litres ▪ 0.5 litre

16. Convert 1.2 m³ to cm³.
a *Don't include the units in your answer.*
b
c ▪ 1,200,000 ▪ 1200000

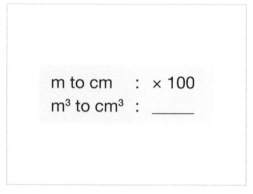

m to cm   : × 100
m³ to cm³ : _____

17. What is the volume of the cuboid in cm³?
a *Don't include the units in your answer.*
b
c ▪ 40,000 ▪ 40000

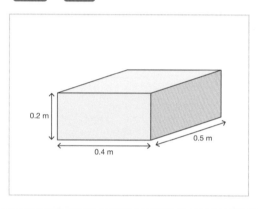

0.2 m

0.4 m

0.5 m

Level 3: Reasoning - With metric units of volume.

✱ **Required:** 5/5   ✱ **Student Navigation:** on
✱ **Randomised:** off

18. To convert from m³ to mm³ you need to multiply
a by _____.
b
c ▪ 1,000,000,000 ▪ 1000000000

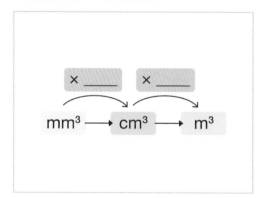

× _____   × _____

mm³ ⟶ cm³ ⟶ m³

Level 3: *cont.*

19. Calculate the volume of the triangular prism in cm³.

Don't include the units in your answer.

▪ 2.16

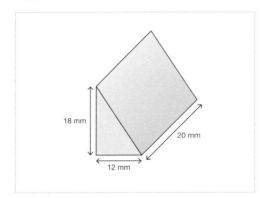

20. Carlos says, "100 cm = 1 m, so 100 cm³ = 1 m³". Is Carlos correct? Explain your answer.

21. Select the symbol that makes the following statement true:

892,400 cm³ __ 0.9 m³

1/3 ▪ < ▪ = ▪ >

22. Sofia says, "My bottle of shower gel holds 240 ml of liquid. This is the same as 240 cm³." Is Sofia correct? Explain your answer.

Level 4: Problem Solving - Complex problems involving metric units of volume.

✱ **Required:** 5/5 ✱ **Student Navigation:** on
✱ **Randomised:** off

23. If a cube has an volume of 0.512 m³, what is the length of one side in cm?

Include the units cm (centimetres) in your answer.

▪ 80 cm ▪ 80 centimetres

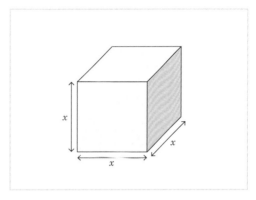

24. Calculate the volume of the composite shape in cm³.

Don't include the units in your answer.

▪ 13,500,000 ▪ 13500000

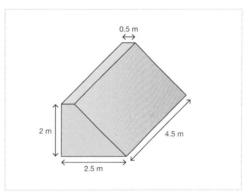

25. Stacey buys a new paddling pool. How many litres of water will it take to fill the paddling pool?

Include the units l (litres) in your answer.

▪ 120 l ▪ 120 litres

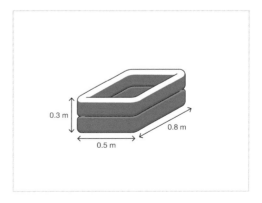

26. The area of each face of a cube is 1,225 mm².
 Calculate the volume of the cube to the nearest
cubic centimetre.
Don't include the units in your answer.

▪ 43

27. How many cartons will 8 litres of lemonade fill?

 ▪ 20

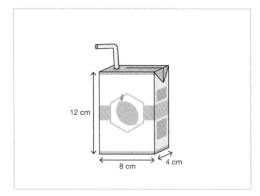

12 cm

8 cm   4 cm

Calculate and Solve Problems Involving Perimeters of Composite Shapes

Competency: Calculate and solve problems involving: perimeters of 2D shapes (including circles), areas of circles and composite shapes.

Quick Search Ref: 10307

Correct: Correct.   Wrong: Incorrect, try again.   Open: Thank you.

Level 1: Understanding - Perimeter of composite shapes.

✱ **Required:** 10/10   ✱ **Student Navigation:** on   ✱ **Randomised:** off

1. Square A has an area of 64 square centimetres
   a
   b (cm²) and square B has an area of 25cm². What is
   c the perimeter of the composite shape?
   Include the units cm (centimetres) in your answer.

   ▪ **72 centimetres** ▪ **72 cm**

   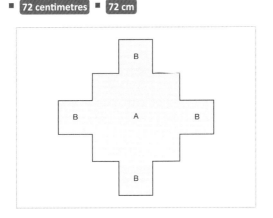

2. What is the perimeter of shape A?
   a
   b *Include the units cm (centimetres) in your answer.*
   c
   ▪ **38 centimetres** ▪ **38 cm**

   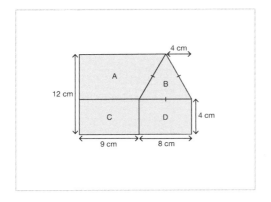

3. The square-shaped swimming pool has a width of
   a 3 metres (m) and the lawn has an area of 45
   b square metres (m²).
   c If the length and width of the garden both
   measure under 10 metres (m) and are whole
   numbers, what is the perimeter of the garden?
   Include the units m (metres) in your answer.

   ▪ **30 m** ▪ **30 metres**

   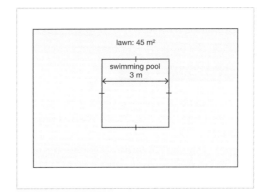

4. The playground at Hamilton School is made up of a
   a square and two equilateral triangles. What is the
   b perimeter of the playground?
   c *Include the units m (metres) in your answer.*

   ▪ **54 m** ▪ **54 metres**

   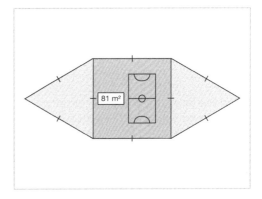

5. A composite shape is made from an equilateral
a triangle and three identical squares. Calculate the
b perimeter of the composite shape in terms of *x*.
c *Give your answer in its simplest form*.

▪ 12x - 3

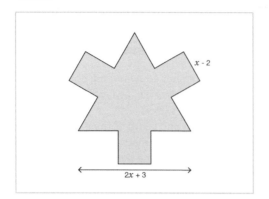

6. An equilateral triangle has an area of 18 square
a centimetres (cm²) and perpendicular height of 3
b centimetres (cm). What is its perimeter?
c *Include the units cm (centimetres) in your answer*.

▪ 36 centimetres ▪ 36 cm

7. Darryl needs to buy new skirting boards for his
a master bedroom. How many metres (m) should he
b order?
c *Include the units m (metres) in your answer*.

▪ 30.8 metres ▪ 30.8 m

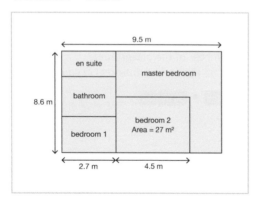

8. Lilly walks from the church (A) to the train station
a (B). How far does she walk in total?
b *Include the units m (metres) in your answer*.
c

▪ 36 m ▪ 36 metres

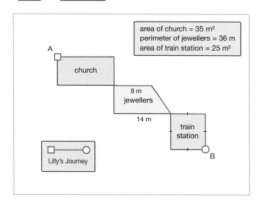

9. A composite shape is made from four identical
a rectangles and a larger rectangle. Calculate the
b perimeter of the composite shape in terms of *x*.
c *Give your answer in its simplest form*.

▪ 14x + 26

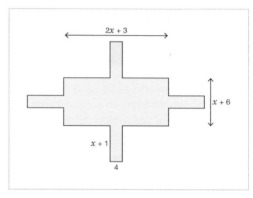

10. The vertices of the rhombus align with the
a midpoints in the length and width of the
b rectangle.
c If the perimeter of each triangle is 21.2
centimetres (cm), what is the perimeter of the
rhombus?
Include the units cm (centimetres) in your answer.

▪ 35.2 cm ▪ 35.2 centimetres

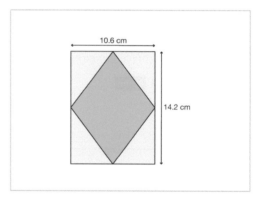

Calculate and Solve Problems Involving the Area of Composite Shapes

Competency: Calculate and solve problems involving: perimeters of 2D shapes (including circles), areas of circles and composite shapes.

Quick Search Ref: 10252

Correct: Correct.   **Wrong:** Incorrect, try again.   **Open:** Thank you.

Level 1: Problem Solving - Calculating the area of compound shapes.

⚙ Required: 10/10   **⚙ Student Navigation:** on   **⚙ Randomised:** off

1. Harry's grandad has a square vegetable patch and nine square flower beds in his garden. In square metres (m²), what is the area of the lawn where he grows apples?
   Don't include the units in your answer.

   ▪ 27

   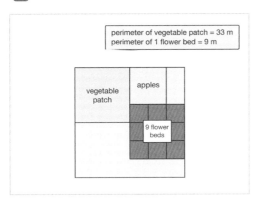

2. The area of the shaded cuboid face is _____ cm²
   Calculate the area of the shaded cuboid face in square centimetres (cm²).
   Don't include the units in your answer.

   ▪ 90

   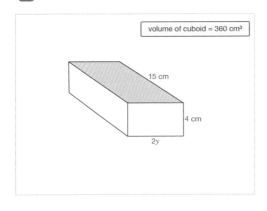

3. A family is tiling their kitchen and utility room floors. If tiles cost £12.50 per square metre (m²), how much will the tiles cost in total?
   Include the £ sign in your answer.

   ▪ £418.75

   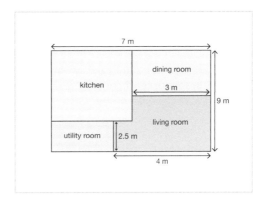

4. Riley is making a flag of the Bahamas for a parade. What is the minimum amount of yellow material he needs?
   Don't include the units in your answer.

   ▪ 68

   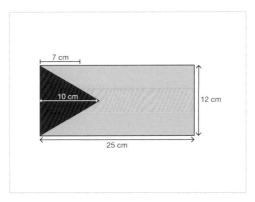

5. Calculate the purple shaded area of the right-angled trapezium.
Don't include the units in your answer.

▪ 743

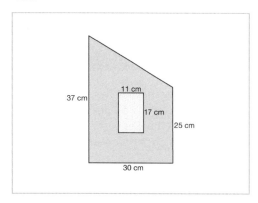

6. Triangle A has a perimeter of 57 centimetres (cm) and triangle B has a perimeter of 39 centimetres. If both are equilateral triangles, what is the area of shape C in square centimetres (cm²)?
Don't include the units in your answer.

▪ 247

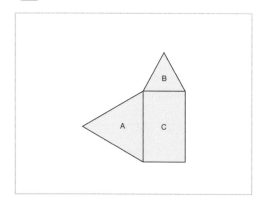

7. If turf costs £3.50 per square metre (m²), how much will it cost to returf Jamie's lawn?
Include the £ sign in your answer.

▪ £176.75

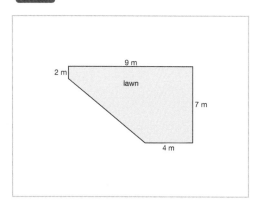

8. Some friends are making a Trinidad and Tobago flag for a school project. The ratio of black:white is 3:2. How much white card do they need in square cm (cm²)?
Don't include units in your answer.

▪ 150

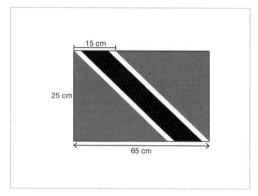

9. The arrowhead is made from a rectangle and an equilateral triangle. What is the area of the arrowhead in terms of x?
Give your answer in its simplest form.

▪ 13x + 6   ▪ 6 + 13x

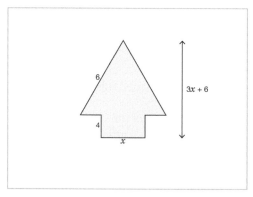

10. The area of the large square is 169 cm². The three smaller squares each have an area of x^2. The areas where the smaller squares overlap measure 1 cm². What is the value of x?

▪ 5

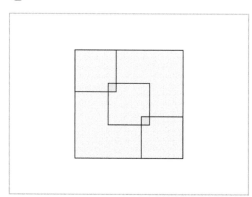

Calculate the Area of a Circle

Competency: Calculate and solve problems involving: perimeters of 2D shapes (including circles), areas of circles and composite shapes.

Quick Search Ref: 10121

Correct: Correct.   Wrong: Incorrect, try again.   Open: Thank you.

Level 1: Understanding - Calculating the area of a circle.

🌼 **Required:** 7/10   🌼 **Student Navigation:** on   🌼 **Randomised:** off

---

1. Select the correct formula to calculate the area of a circle.

▪ πr^2 ▪ $2\pi r$ ▪ πd

1/3

2. The area of the circle = __ π cm².

  ▪ 64

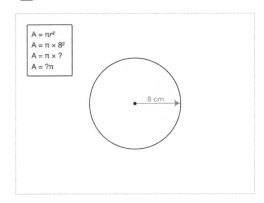

3. The area of the circle = __ π m².

  ▪ 25

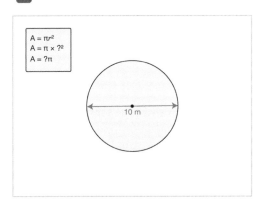

4. The area of the circle is _____ m².

  ▪ 28

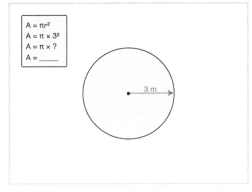

5. The area of the circle is _____ cm².
Give your answer to the nearest whole number.

  ▪ 113

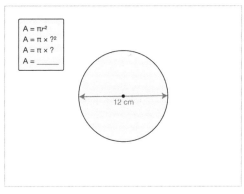

6. The area of the circle is _____ cm²
Give your answer to the nearest whole number.

  ▪ 55

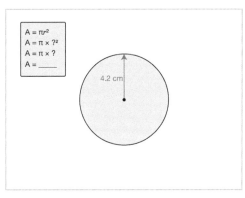

7. The area of the circle is _____ cm²
Give your answer to the nearest whole number.

  ▪ 38

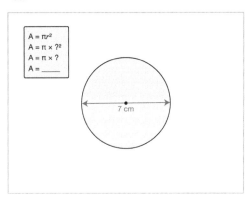

8. The area of the circle is_____ mm²

 ▪ 133

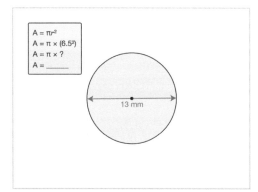

$A = \pi r^2$
$A = \pi \times (6.5^2)$
$A = \pi \times ?$
$A = $_____

13 mm

9. The area of the circle is_____ cm²

 ▪ 70.56

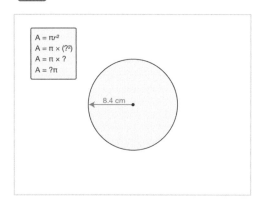

$A = \pi r^2$
$A = \pi \times (?^2)$
$A = \pi \times ?$
$A = ?\pi$

8.4 cm

10. The area of the circle is_____ m²

 ▪ 227

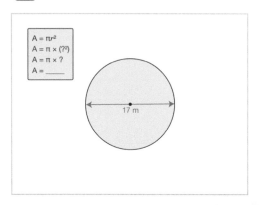

$A = \pi r^2$
$A = \pi \times (?^2)$
$A = \pi \times ?$
$A = $_____

17 m

Level 2: Fluency - Calculating the area of a circle in context rounding to a specified number of decimal places and significant figures.

✸ **Required:** 7/10   ✸ **Student Navigation:** on
✸ **Randomised:** off

11. A DVD has a radius of 6 centimetres (cm). What is the area of a DVD in cm²? Give your answer to 3 decimal places.
Don't include the units in your answer.

▪ 113.097

12. What is the difference in area (cm²) between the two circles? Give your answer to 2 decimal places. *Don't include the units in your answer.*

 ▪ 21.21

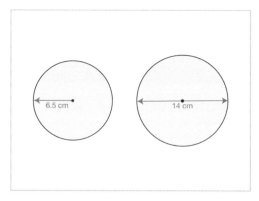

6.5 cm   14 cm

13. If Jamie's paddling pool has a diameter of 2.3 metres (m), what is its area in square metres (m²)? Give your answer to 3 significant figures.
Don't include the units in your answer.

 ▪ 4.15

14. The area of the semicircle is_____ cm²
Give your answer to 1 d.p.

 ▪ 66.4

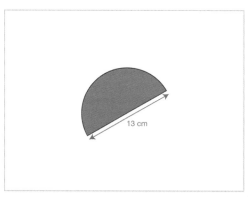

13 cm

15. The two circles share the same centre. What is the area of the shaded ring in square centimetres (cm²)? Give your answer to 1 decimal place.
Don't include the units in your answer.

 ▪ 241.9

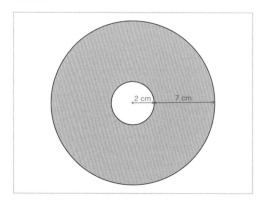

2 cm   7 cm

16. The area of the 3/4 circle is_____ cm²
Give your answer to 4 significant figures.

▪ 350.7

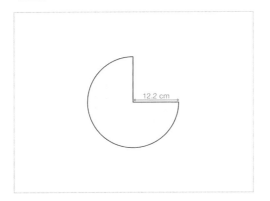

17. What is the area of the shaded shape in square
centimetres (cm²)? Give your answer to 2 decimal
places.
Don't include the units in your answer.

▪ 103.87

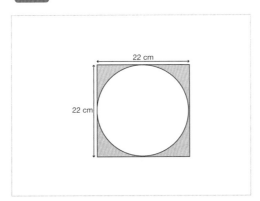

18. Mrs Conan wears a necklace with a pendant. What
is the area of the pendant in square centimetres
(cm²)? Give your answer to 4 significant figures.
Don't include the units in your answer.

▪ 1.539

19. What is the difference in area (cm²) between the
two circles? Give your answer to 2 decimal places.
Don't include the units in your answer.

▪ 29.03

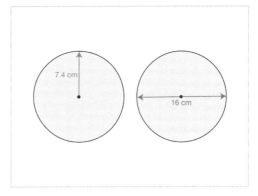

20. Calculate the area of the quadrant in square
metres (m²). Give your answer to 3 significant
figures.
Don't include the units in your answer.

▪ 177

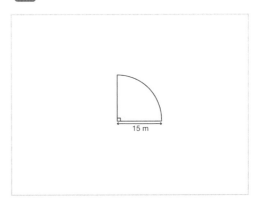

Level 3: Reasoning - Inverse and proof involving the
area of a circle.

✿ **Required:** 5/5   ✿ **Student Navigation:** on
✿ **Randomised:** off

21. If you double the radius of a circle, what happens
to the area? Use proof to explain your answer.

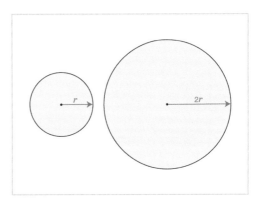

Level 3: *cont.*

22. The area of the circle is 28.27 square metres (m²).
a
b Calculate the radius of the circle to the nearest
c whole number.
Include the units m (metres) in your answer.

▪ 3 metres ▪ 3 m

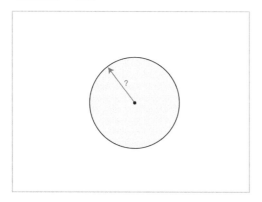

23. Billy says he knows which circle has the largest
a area without calculating. Is this possible? Explain
b your answer.
c

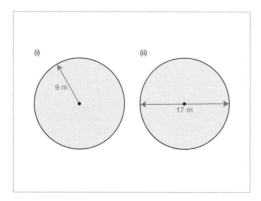

24. Calculate the volume of the cylinder in cubic
a centimetres (cm³).
b *Don't include the units in your answer.*
c

▪ 3386.6 ▪ 3,386.6

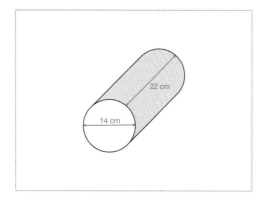

25. Prove that the area of a circle is less than 4r².
a
b
c

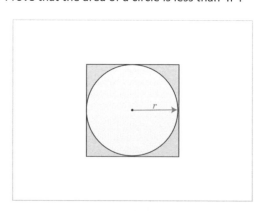

Level 4: Problem Solving - Calculating the area of
circles in complex problems.

✱ **Required:** 5/5   ✱ **Student Navigation:** on
✱ **Randomised:** off

26. Andy made a circular rug with materials costing
a £4.75 per square metre. If he spent £20 on
b materials, what is the diameter of Andy's rug?
c Give your answer to 2 decimal places.
Include the units m (metres) in your answer.

▪ 2.32 m ▪ 2.32 metres

27. Calculate the surface area of the cylinder in square
a metres (m²). Give your answer to 2 decimal places.
b *Don't include the units in your answer.*
c

▪ 2883.98 ▪ 2,883.98

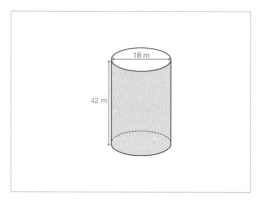

28. There are 77 stickers on an A4 sheet. Each sticker
a has a radius of 1.3 cm. What percentage of the
b sheet is not covered by stickers?
c Give your answer to the nearest whole percent.
Include the % sign in your answer.

▪ 34%

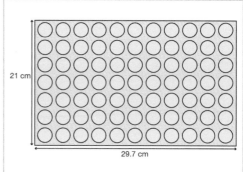

29. Sasha's patio is formed by three circles with the
ratio 2:3:5. If patio B has a diameter of 4 metres
(m), what is the area of the whole patio in metres?
Give your answer to 3 significant figures.
Don't include the units in your answer.

a
b
c

▪ 41.9

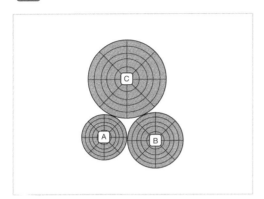

30. Calculate the area of the circle as a percentage of
the area of the square.
Give your answer to 1 decimal place.
Include the % sign in your answer.

a
b
c

▪ 78.5%

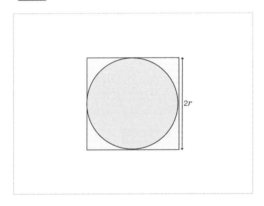

Calculate the Circumference of a Circle

Competency: Calculate and solve problems involving: perimeter of 2-D shapes (including circles), area of circles and composite shapes.

Quick Search Ref: 10214

Correct: Correct.   Wrong: Incorrect, try again.   Open: Thank you.

Level 1: Understanding - Parts of a circle, pi and the formula for the circumference.

✴ **Required:** 7/10   ✴ **Student Navigation:** on   ✴ **Randomised:** off

1. Arrange the lengths in the following order:
   ↑↓ Radius
   Diameter
   Circumference

   ▪ 15 cm ▪ 30 cm ▪ 94.25 cm

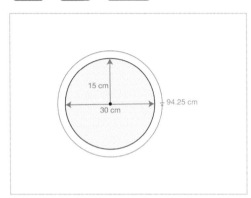

2. What is the radius of the circle in metres (m)?
   123 *Don't include the units in your answer.*

   ▪ 21

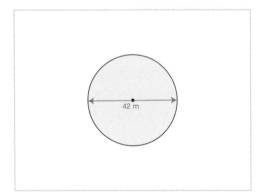

3. If the circumference of a circle is approximately 3 times longer than the diameter, what is the approximate circumference of the circle in millimetres (mm)?
   123 *Don't include the units in your answer.*

   ▪ 93

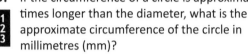

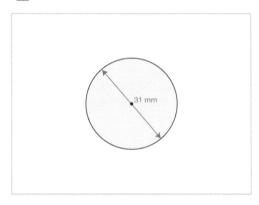

4. The exact circumference of a circle = π × diameter (*C* = π*d*). π is an irrational number, so which option shows the closest approximation of π?

   1/3 ▪ **3.14** ▪ **22/7** ▪ 3.14159

5. Use your calculator to find the circumference of the circle. Give your answer to the nearest whole millimetre (mm).
   123 *Don't include the units in your answer.*

   ▪ 85

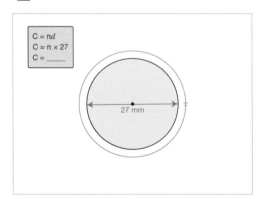

6. Calculate the circumference of the circle to the nearest whole metre (m).
Don't include the units in your answer.

▪ **75**

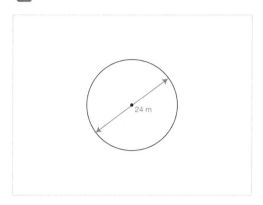

7. Calculate the circumference of the circle to the nearest whole centimetre (cm).
Don't include the units in your answer.

▪ **50**

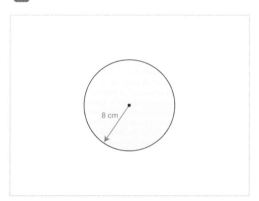

8. The circumference of a circle is approximately 3 times its diameter. What is the approximate circumference of the circle in millimetres (mm)?
Don't include the units in your answer.

▪ **156**

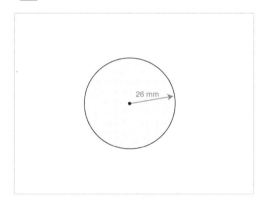

9. Calculate the circumference of the circle to the nearest whole metre (m).
Don't include the units in your answer.

▪ **25**

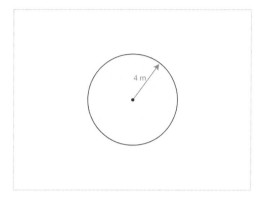

10. Calculate the circumference of the circle to the nearest whole centimetre (cm).
Don't include the units in your answer.

▪ **119**

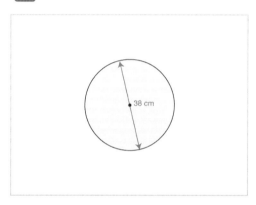

Level 2: Fluency - Calculating the circumference of a circle.

✿ **Required: 7/10**   ✿ **Student Navigation:** off
✿ **Randomised:** off

11. What is the circumference of the circle rounded to 2 decimal places?
Include the units mm (millimetres) in your answer.

▪ **67.23 mm** ▪ **67.23 millimetres**

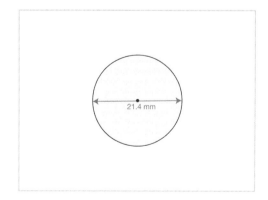

12. The circumference of the circle = _ π mm.

   ▪ 29.6

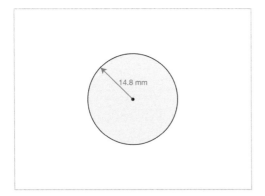

14.8 mm

13. What is the length of the arc of the quadrant? Give your answer to 3 significant figures.
Include the units mm (millimetres) in your answer.

▪ 14.5 millimetres ▪ 14.5 mm

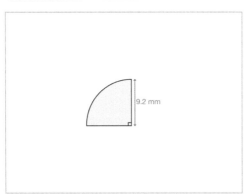

9.2 mm

14. What is the radius of the circle to 4 significant figures?
Include the units m (metres) in your answer.

▪ 3.008 metres ▪ 3.008 m

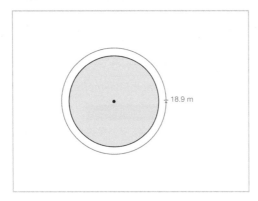

18.9 m

15. Calculate the perimeter of the three-quarter circle rounded to 1 decimal place.
Include the units cm (centimetres) in your answer.

▪ 18.8 centimetres ▪ 18.8 cm

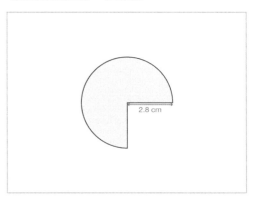

2.8 cm

16. Daisy skips around her circular patio 3 times and travels a distance of 32 metres (m). What is the diameter of her patio to 1 decimal place?
Include the units m (metres) in your answer.

▪ 3.4 metres ▪ 3.4 m

17. Calculate the circumference of the circle in terms of *t*.

  ▪ C = tπ ▪ C = π(2t) ▪ C = π(t²)

1/3

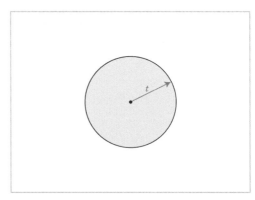

t

18. Calculate the perimeter of the semicircle to 1 decimal place.
Include the units m (metres) in your answer.

▪ 18.5 metres ▪ 18.5 m

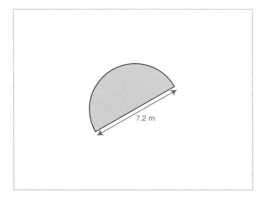

7.2 m

Level 2: *cont.*

19. Calculate the diameter of the circle and give your
answer to 2 decimal places.
Include the units m (metres) in your answer.

a
b
c

■ 16.49 m   ■ 16.49 metres

20. 1,500 cm (centimetres) of thread is wound around
a cotton reel 120 times. What is the diameter of
the cotton reel to the nearest whole centimetre?
Include the units cm (centimetres) in your answer.

a
b
c

■ 4 centimetres   ■ 4 cm

Level 3:   Reasoning - Including proof and comparison.

🌼 **Required:** 5/5   🌼 **Student Navigation:** on
🌼 **Randomised:** off

21. Jonny says he can find the circumference of a
circle by using the formula 2πr. Is he correct?
Explain your answer.
π can be written as pi.

a
b
c

22. Which shape has the largest perimeter?

□
☒
□

1/3

■ square   ■ triangle   ■ circle

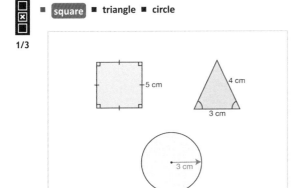

23. The two small circles have a combined
circumference equal to that of the large circle.
What is the radius of one of the small circles?
Include the units cm (centimetres) in your answer.

a
b
c

■ 2 centimetres   ■ 2 cm

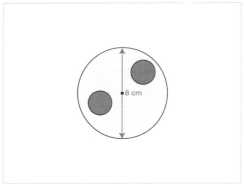

24. Prove that the circumference of a circle must be
less than 4*d*.
π can be written as pi.

a
b
c

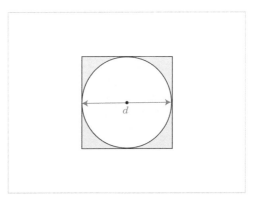

25. An 8-inch pizza is cut into six equal pieces. What is
the perimeter of one slice to the nearest inch?
Don't include the units in your answer.

1
2
3

■ 12

Level 4:   Problem Solving - Problems involving the
circumferences of circles.

🌼 **Required:** 5/5   🌼 **Student Navigation:** on
🌼 **Randomised:** off

26. A bicycle wheel has a diameter of 17.5 cm. How
many full turns does the wheel make over a 4
kilometre (km) distance?

a
b
c

■ 7,275   ■ 7275

27. Julia is digging out three identical overlapping
circles to create a flowerbed. She wants to
surround her flowerbed with a brick border. If
every brick is 15 centimetres (cm) in length, how
many bricks will she need?

▪ 92

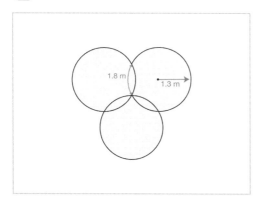

28. The London Eye has a radius of 60 metres (m). If it
takes 30 minutes for a capsule to perform a full
rotation, how far would you travel if you stayed on
the London Eye for 3 hours? Give your answer to 2
decimal places.
Include the units km (kilometres) in your answer.

▪ 2.26 kilometres ▪ 2.26 km

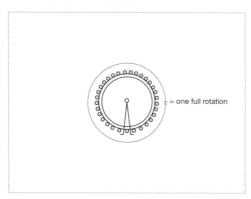

29. Form the algebraic expression for the perimeter of
the composite shape, which is made from a
square, a circle and a triangle.
Give your answer in its simplest form.

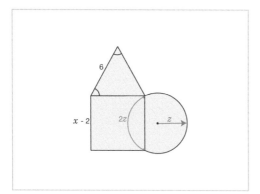

30. A piece of A4 paper measures 21 cm × 29.7 cm
and can be rolled two different ways to make a
cylinder. What is the difference between the radii
of the two cylinders? Give your answer to 1
decimal place.
Include the units cm (centimetres) in your answer.

▪ 1.4 centimetres ▪ 1.4 cm

Use the Formula to Calculate the Area of a Trapezium

Competency: Derive and apply formulae to calculate and solve problems involving: perimeter and area of triangles, parallelograms, trapezia, volume of cuboids (including cubes) and other prisms (including cylinders).

Quick Search Ref: 10253

Correct: Correct.   **Wrong:** Incorrect, try again.   **Open:** Thank you.

Level 1: Understanding - Recognising trapezia and deriving the formula for the area.

✿ **Required:** 7/10   ✿ **Student Navigation:** on   ✿ **Randomised:** off

1. Which two statements describe a trapezium?

    2/5
    - ■ **Has one pair of parallel lines.**
    - ■ Has at least one right angle.
    - ■ All angles are congruent (the same).
    - ■ **Is a type of quadrilateral.**
    - ■ All sides are congruent (the same).

    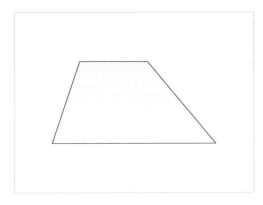

2. Select the three shapes that are trapezia.

    3/6
    ■ **(i)** ■ (ii) ■ **(iii)** ■ **(iv)** ■ (v) ■ (vi)

    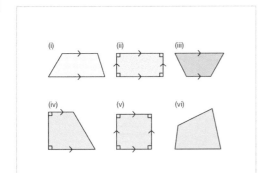

3. What shape is made by the two congruent trapezia?

    1/4
    ■ Rectangle ■ Square ■ **Parallelogram** ■ Trapezium

    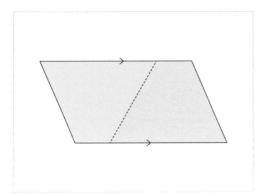

4. The area of the parallelogram is _____ m².

    1
    2
    3
    ■ 12

    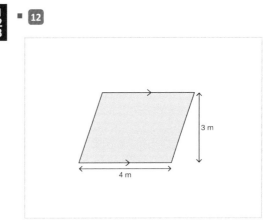

5. Use the area of the parallelogram to calculate the area of the shaded trapezium.
    Don't include the units in your answer.

    1
    2
    3
    ■ 6

    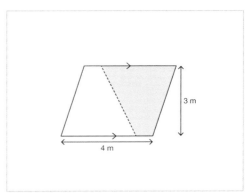

6. If both trapezia are congruent, the area of the shaded trapezium is _____ m².

    1
    2
    3
    ■ 16

    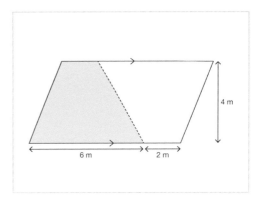

Level 1: *cont.*

7. Use your understanding of parallelograms
    to identify the formula for the area of a trapezium.

    ■ (2a + 2b) × h   ■ ½(a + b) × h   ■ a × b × h

    1/3

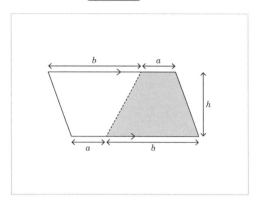

8. Use the area of the parallelogram to calculate the
    area of one trapezium (both are congruent) in m².

    ■ 24

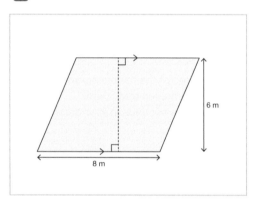

9. If both trapezia are congruent, the area of one
    trapezium is ____ m².

    ■ 8

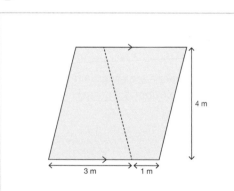

10. What is the formula for the area of one
    trapezium?

    ■ (2a + 2b) × h   ■ (a + b) × h   ■ ½(a + b) × h

    1/3

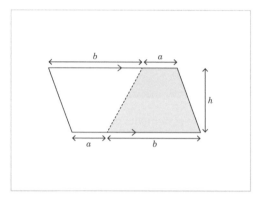

Level 2:   Fluency - Calculating the area and comparing
               trapezia.

✴ **Required:** 7/10   ✴ **Student Navigation:** on
✴ **Randomised:** off

11. The area of the trapezium is _____ cm².

    ■ 13.5

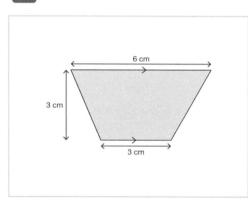

12. The area of the trapezium is _____ cm².

    ■ 55

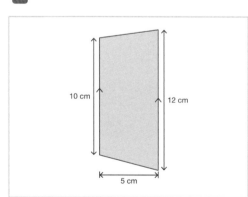

13. The area of the trapezium is _____ m².

  ▪ 24.75

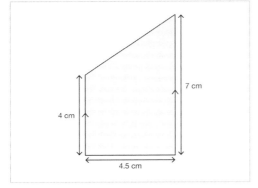

14. Which trapezium has the greatest area?

 ▪ (i) ▪ (ii) ▪ (iii)

1/3

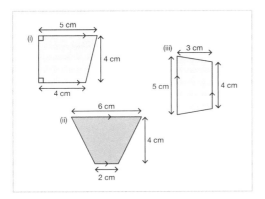

15. Which two trapezia have the same area?

 ▪ (i) ▪ (ii) ▪ (iii) ▪ (iv)

2/4

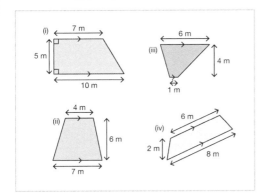

16. Arrange the trapezia by area in ascending order (smallest first).

 ▪ (i) ▪ (iv) ▪ (ii) ▪ (iii)

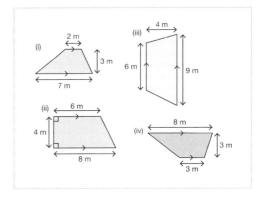

17. According to the plan, what is the area of the bathroom?
Don't include the units in your answer.

 ▪ 17.63

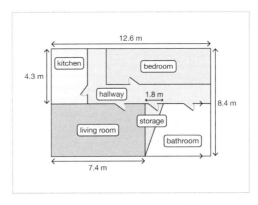

18. The area of the trapezium is _____ m².

 ▪ 57

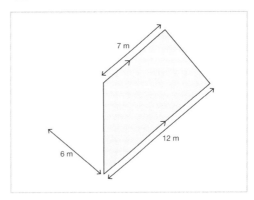

19. Which trapezium has the smallest area?

 ▪ (i) ▪ (ii) ▪ (iii)

1/3

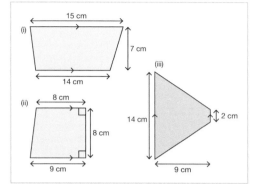

20. Arrange the trapezia in descending order of size by area (largest first).

▪ (ii) ▪ (i) ▪ (iv) ▪ (iii)

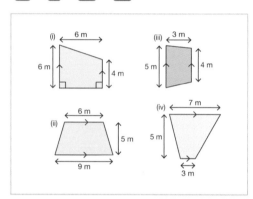

Level 3: Reasoning - Proof and inverse operations.

✱ **Required:** 5/5   ✱ **Student Navigation:** on
✱ **Randomised:** off

21. What is the area of the trapezium in terms of *r*, *s*
and *t*?

▪ ½(r × s × t) ▪ ½(r × s) × t ▪ ½(r + s) × t ▪ ½(t + r + s)

1/4

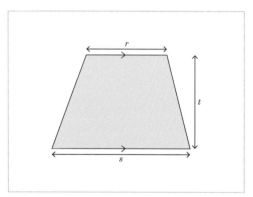

22. Calculate the perpendicular height of the
trapezium.
Include the units cm (centimetres) in your answer.

▪ 8 cm ▪ 8 centimetres

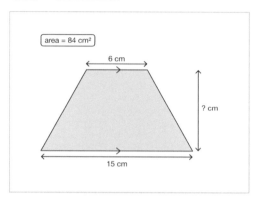

23. Amelia is working out the area of the L-shape by
calculating the area of the two trapezia that it's
formed by. Ben says there is another way to
calculate the area. Is he correct? Explain your
answer.

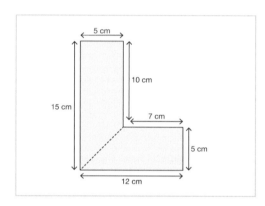

24. Which two trapezia have the same area?

▪ (i) ▪ (ii) ▪ (iii) ▪ (iv) ▪ (v) ▪ (vi)

2/6

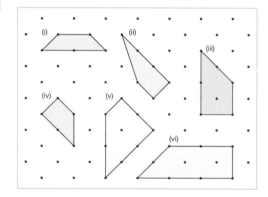

Level 3: *cont.*

25. Prove that the area of a trapezium = ½(a + b) × h.

a
b
c

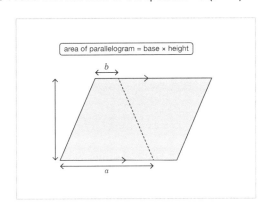

Level 4: Problem Solving - Context problems involving the area of trapezia.

✿ **Required:** 5/5   ✿ **Student Navigation:** on
✿ **Randomised:** off

26. The area of the shaded trapezium is 56 cm².
a Calculate the perimeter of the **shaded** trapezium.
b *Include the units cm (centimetres) in your answer.*
c

▪ 41 cm ▪ 41 centimetres

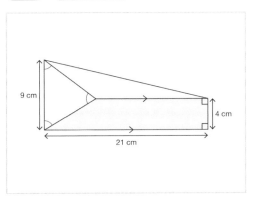

27. Find the area of the **shaded** shape.
a *Don't include the units in your answer.*
b
c ▪ 3,270 ▪ 3270

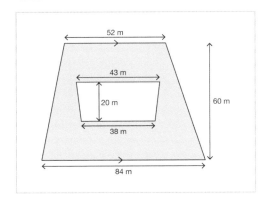

28. Bobby has drawn a house using 2D shapes.
1 Calculate its area in cm².
2 *Don't include the units in your answer.*
3

▪ 483

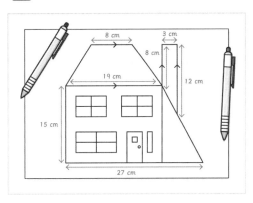

29. Number 14 and number 16 are two identical
a houses (each 3 metres wide).
b The widest part of number 14's garden is twice as
c wide as number 16's.
The total area of the two gardens is 128.25 m².
Calculate the length of the fence that separates
the two gardens.
Include the units m (metres) in your answer.

▪ 9.5 metres ▪ 9.5 m

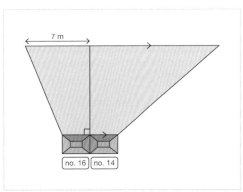

30. Colin wants to returf his lawn. Each roll of turf
a covers an area of 3 square metres (m²). If a roll of
b turf costs £9.42 how much will it cost to returf the
c whole lawn?
Include the £ sign in your answer.

▪ £329.70

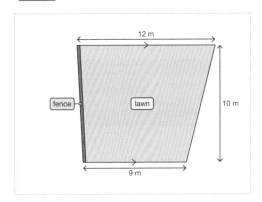

Calculate the Surface Area of Prisms, Including Cylinders

Competency: Derive and apply formulae to calculate and solve problems involving: perimeter and area of triangles, parallelograms, trapezia, volume of cuboids (including cubes) and other prisms (including cylinders).

Quick Search Ref: 10066

Correct: Correct.   Wrong: Incorrect, try again.   Open: Thank you.

Level 1: Understanding - Calculating the area of faces of prisms.

✿ **Required:** 7/10   ✿ **Student Navigation:** on   ✿ **Randomised:** off

1. The area of the square is _____ cm².

 ▪ 9

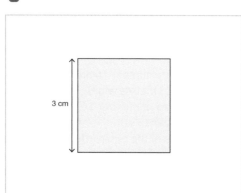

2. What is the definition of surface area?

▪ The amount of 3-dimensional space that an object fills (measured in cubic units).

1/3 ▪ The total area of the surface of a 3-dimensional object.

▪ The size of a surface or the measure of the space inside a 2-dimensional shape.

3. The surface area of the cube is _____ cm².

 ▪ 54

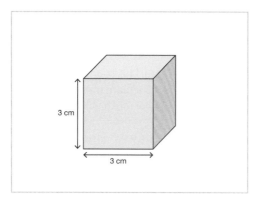

4. Select the three properties of a prism.

▪ The cross section is the same all along its length.

▪ It is a 2D shape. ▪ It is a 3D shape.

3/5 ▪ It is a type of pyramid

▪ It has at least two identical faces.

5. Identify the net for each of the following and arrange the options in the same order:
Cube
Triangular prism
Pentagonal prism
Cylinder
Trapezium prism

▪ (i) ▪ (ii) ▪ (v) ▪ (iii) ▪ (iv)

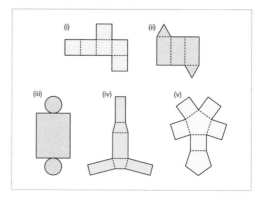

6. The surface area of the cuboid is _____ cm².

 ▪ 148

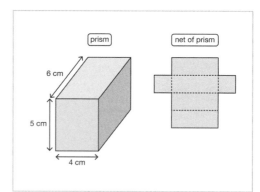

Level 1: cont.

7. The surface area of the cuboid is _____cm².

  ▪ 484

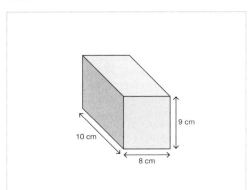

8. Identify the net for each of the following prisms and arrange the options in the same order:
Cube
Triangular prism
Pentagonal prism
Irregular hexagonal prism

▪ (ii) ▪ (iv) ▪ (i) ▪ (iii)

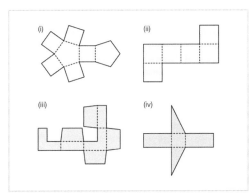

9. The surface area of the cuboid is _____cm².

1 2 3 ▪ 406

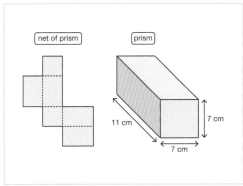

10. The surface area of the cuboid is _____cm².

1 2 3 ▪ 584

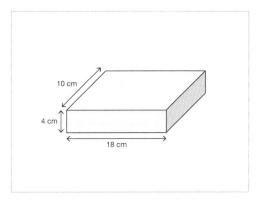

Level 2: Fluency - Calculating the surface area of various prisms and cylinders.

❋ **Required:** 7/10 ❋ **Student Navigation:** on
❋ **Randomised:** off

11. The surface area of the triangular prism is _____cm².

1 2 3 ▪ 294

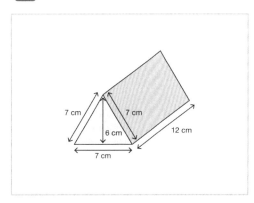

12. The surface area of the hexagonal prism is _____cm².

1 2 3 ▪ 669.9

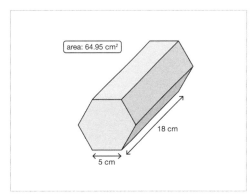

area: 64.95 cm²

13. If the regular pentagonal face has an area of 72
square metres (m²), what is the total surface area
of the prism?
Don't include the units in your answer.

■ 532.2

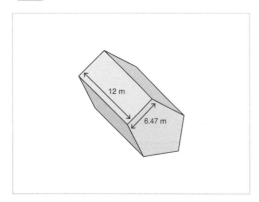

14. The surface area of the trapezium prism is
_____cm².

■ 210.6

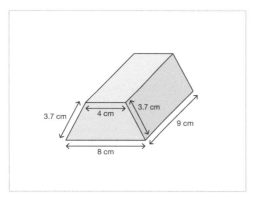

15. Calculate the surface area of the cylinder in square
metres (m²). Give your answer to the nearest
whole number.
Don't include the units in your answer.

■ 283

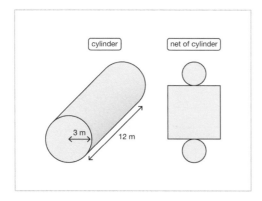

16. The surface area of the hexagonal prism is
_____cm².

■ 312

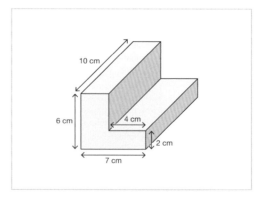

17. The surface area of the triangular prism is
_____m².

■ 364

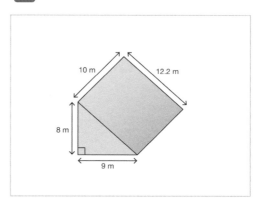

18. If the regular heptagonal face has an area of 14.54
square centimetres (cm²), what is the total surface
area of the prism?
Don't include the units in your answer.

■ 211.08

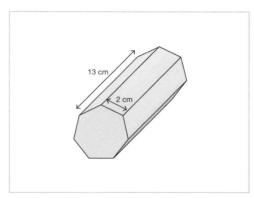

Level 2: *cont.*

19. The surface area of the pentagonal prism is

_____cm².

■ 263.6

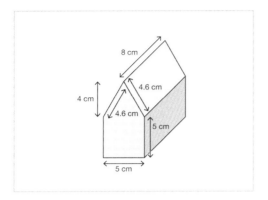

20. The surface area of the trapezium prism is
_____cm².

■ 255.6

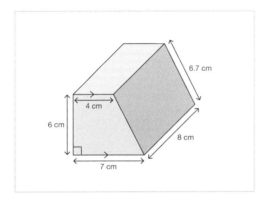

Level 3: Reasoning - Surface area of prisms and cylinders (including algebra).

✸ **Required:** 5/5 ✸ **Student Navigation:** on
✸ **Randomised:** off

21. Each of the prisms has the same cross-sectional
area and length.
a
b
c
Kayla says, "If each triangle has the same area, then the sum of their side lengths must be equal, and so the surface area will also be equal".
Is Kayla correct? Explain your answer.

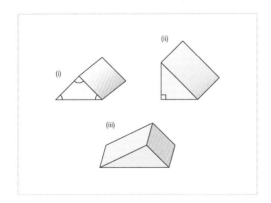

22. Mila makes a mistake when calculating the surface area of the triangular prism shown. What number has she forgotten to add?

■ 24

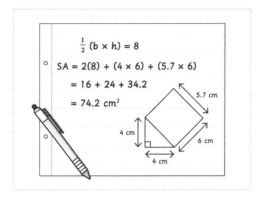

$$\frac{1}{2}(b \times h) = 8$$
$$SA = 2(8) + (4 \times 6) + (5.7 \times 6)$$
$$= 16 + 24 + 34.2$$
$$= 74.2 \ cm^2$$

23. A piece of A4 paper can be rolled in two different ways to make two different cylinders. If the two cylinders were solid, would they have the same surface area? Explain your answer.
a
b
c

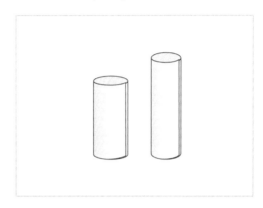

24. What is the surface area of the cuboid in terms of a, b and c?.
a
b
c
Give your answer in its simplest form.

■ 2(bc + ac + ab) ■ 2(ab + ac + bc) ■ 2(ab + bc + ac)
■ 2(bc + ab + ac) ■ 2(ac + bc + ab) ■ 2(ac + ab + bc)

Level 3: *cont.*

25. The surface area of the pyramid is _____cm².

 ▪ 224

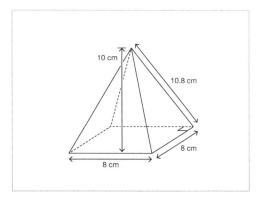

Level 4: Problem Solving - Calculating the surface area of complex prisms and cylinders.

❋ **Required:** 5/5   ❋ **Student Navigation:** on
❋ **Randomised:** off

26. The surface area of the semi-cylinder is _____mm².
Round your answer to 2 d.p.

 ▪ 1,233.67 ▪ 1233.67

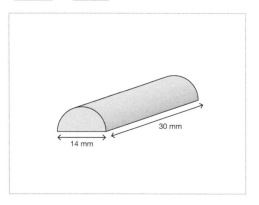

27. The shed has two identical windows and one door. How much paint is needed to cover all of the exterior walls and roof?
Don't include the units in your answer.

▪ 31.48

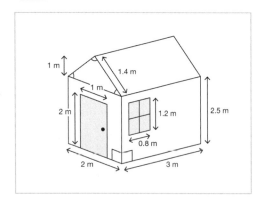

28. Mrs Calvert uses material that costs £6.85 per square metre to cover her chair. The material can only be bought by the square metre. How much does it cost to cover her chair?
Include the £ sign in your answer.

▪ £41.10

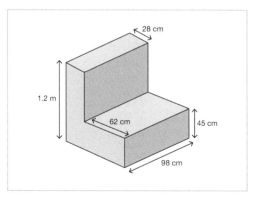

29. The label covers ¾ of the curved surface of the can and has a 2 centimetre (cm) overlap. What is the area of the label to the nearest whole square centimetre (cm²)?
Don't include the units in your answer.

▪ 196

30. This cuboid has a surface area of 377 square metres (m²). The areas of the sides are in the ratio 2:5:6. What is the value of *a*?
Include the units m (metres) in your answer.

▪ 5 metres ▪ 5 m

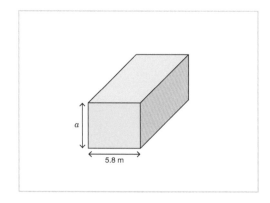

Calculate the Volume of Prisms, Including Cylinders

Competency: Derive and apply formulae to calculate and solve problems involving: perimeter and area of triangles, parallelograms, trapezia, volume of cuboids (including cubes) and other prisms (including cylinders).

Quick Search Ref: 10176

Correct: Correct.   Wrong: Incorrect, try again.   Open: Thank you.

Level 1: Understanding - How to calculate volume of prisms.

🌸 Required: 7/10   🌸 Student Navigation: on   🌸 Randomised: off

---

1. Select the three properties of a prism.

- ■ The cross section is the same all along its length.
- ■ It is a 2D shape. ■ It is a 3D shape.

3/5
- ■ It is a type of pyramid.
- ■ It has at least two identical faces.

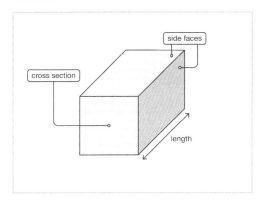

2. What is the definition of volume?

- ■ The amount of space in a container or the amount of liquid that it can hold (measured in non-cubic units).
- ■ The total area of the surface of a 3-dimensional object.

1/3
- ■ The amount of 3-dimensional space that an object fills (measured in cubic units).

3. How many cubes make up the cuboid?

  ■ 18

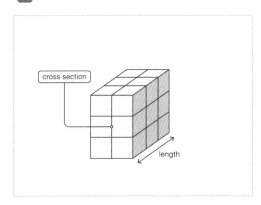

4. The volume of the cuboid is _____ cm³.

  ■ 45

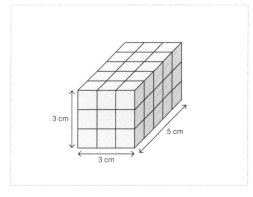

5. The volume of the cuboid is _____ cm³.

  ■ 96

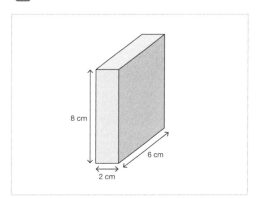

6. The volume of the prism is _____ cm³.

  ■ 24

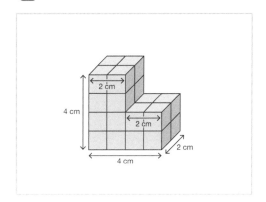

Level 1: *cont*.

7. The volume of the prism is _____cm³.

1
2
3

▪ 84

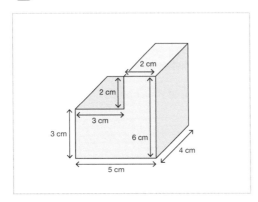

8. The volume of the cuboid is _____cm³.

1
2
3

▪ 32

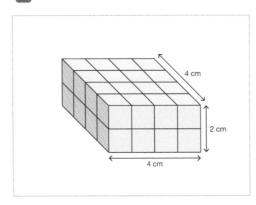

9. The volume of the cuboid is _____cm³.

1
2
3

▪ 60

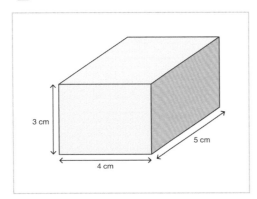

10. The volume of the prism is _____cm³.

1
2
3

▪ 84

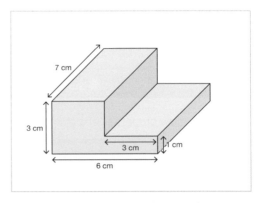

Level 2: Fluency - Calculating the volume of various prisms and a cylinder.

✹ **Required: 7/10**   ✹ **Student Navigation: on**
✹ **Randomised: off**

11. The volume of a prism is found using the formula:

☐
☒
☐

▪ **width × height × length.** ▪ **½(base × height).**
▪ area of cross section × length.

1/3

12. Calculate the volume of the arrow-shaped prism in cubic centimetres (cm³).
Don't include the units in your answer.

1
2
3

▪ 487.62

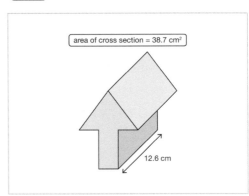

area of cross section = 38.7 cm²

12.6 cm

13. The volume of the cylinder is _____cm³.
Give your answer to 1 decimal place.

a
b
c

▪ 1,357.2 ▪ 1357.2

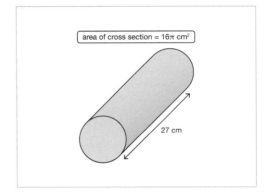

area of cross section = 16π cm²

27 cm

14. If the area of one pentagonal face is 36 cm², what is the volume of the prism in cubic centimetres (cm³)?

Don't include the units in your answer.

▪ 341.28

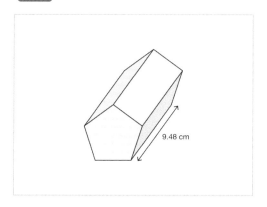

15. The volume of the triangular prism is _____cm³.

▪ 478.8

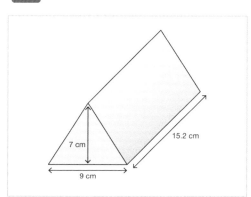

16. The volume of the trapezium is _____cm³.

▪ 205.8

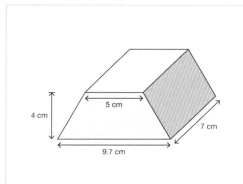

17. The volume of the prism is _____cm³.
Give your answer to 1 decimal place.

▪ 295.1

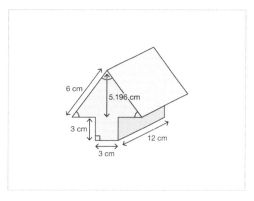

18. The volume of the prism is _____cm³.

▪ 681.8

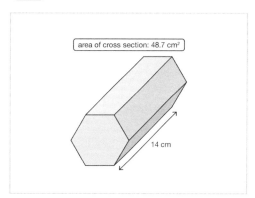

19. The volume of the triangular prism is _____cm³.

▪ 173.25

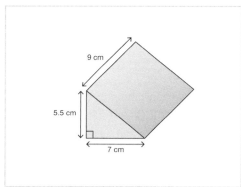

20. The volume of the prism is _____cm³.

▪ 600

Level 3:   Reasoning - Volume of prisms and cylinders.

✻ **Required:** 5/5   ✻ **Student Navigation:** on
✻ **Randomised:** off

21. Poppy says, "I can't find the volume of this cylinder
a because it has no straight edges". Is she correct?
b Explain your answer.
c

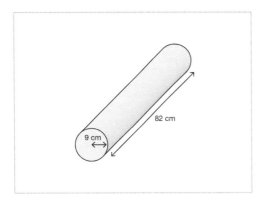

22. Calculate the volume of the semi-cylinder in cubic
a centimetres (cm³).
b Give your answer to 2 decimal places.
c *Don't include the units in your answer.*

▪ 1583.36 ▪ 1,583.36

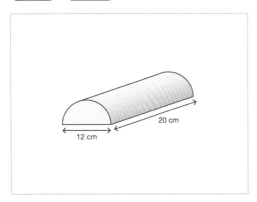

23. This cuboid has a volume of 780.8 cubic
a centimetres (cm³). What is the value of *a*?
b *Include the units cm (centimetres) in your answer.*
c

▪ 8 centimetres ▪ 8 cm

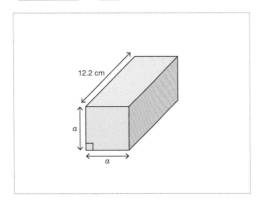

24. Each 3D shape has a cross sectional area of *x* cm²
a and a length of value *y* cm. Isla says, "The shapes
b must all have the same volume". Is she correct?
c Explain your answer.

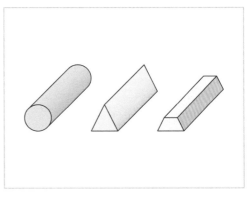

25. Calculate the volume of the prism in cubic metres
1 (m³).
2 *Don't include the units in your answer.*
3

▪ 636

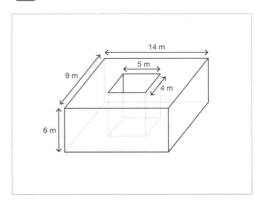

Level 4:   Problem Solving - Volume of prisms and
cylinders.

✻ **Required:** 5/5   ✻ **Student Navigation:** on
✻ **Randomised:** off

26. Calculate the volume of the composite prism.
1
2 ▪ 449.12
3

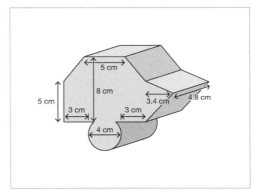

27. A swimming pool is 50 metres (m) long, 25 m
a wide, and 3 m deep. If one cubic metre of water
b costs £1.23, how much will it cost to fill the pool?
c *Include the £ sign in your answer.*

▪ £4612.50 ▪ £4,612.50

Level 4: cont.

28. If the diameter of the pencil lead is 0.7 millimetres (mm), how many cubic centimetres (cm³) of wood does the pencil contain?
Give your answer to 2 decimal places.
Don't include the units in your answer.

▪ 41.04

29. Container (i) is 2/3 full of water and container (ii) is 3/4 full with water. What is the total amount of water in both containers?
Reminder: 1,000 cm³ = 1 litre.
Give your answer to 1 decimal place.
Include the units l (litres) in your answer.

▪ 99.5 litres ▪ 99.5 l

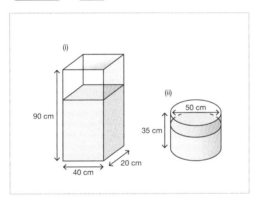

30. The volume of the prism is 1,134 cubic centimetres (cm³). If the measurements *a*, *b* and *c* are in the ratio 2:3:7 and are all whole numbers, what is the area of the cross sectional face in square centimetres (cm²)?
Don't include the units in your answer.

▪ 54

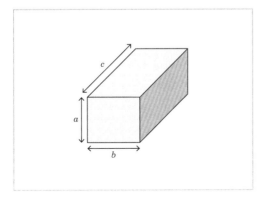

Solve Problems Using Speed

Competency: Use compound units such as speed, unit pricing and density to solve problems.

Quick Search Ref: 10359

Correct: Correct.   Wrong: Incorrect, try again.   Open: Thank you.

Level 1: Understanding - Use proportional reasoning to find missing times and distances.

🏵 Required: 7/10   🏵 Student Navigation: on   🏵 Randomised: off

1. A car travels at a speed of 80 km/h (kilometres per hour). How many kilometres does it travel in 2 hours?
Don't include the unit in your answer.

▪ 160

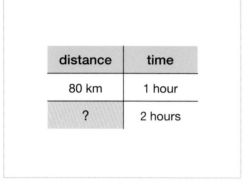

2. A kangaroo runs at 20 m/s (metres per second). How many metres does it travel in 6 seconds?
Don't include the unit in your answer.

▪ 120

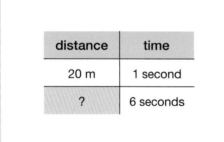

3. A cyclist is riding at a speed of 18 km/h (kilometres per hour). How many kilometres does he travel in half an hour?
Don't include the unit in your answer.

▪ 9

distance	time
18 km	1 hour
?	½ hour

4. A plane flies at 800 km/h (kilometres per hour). How many kilometres does it fly in 0.2 hours?
Don't include the unit in your answer.

▪ 160

distance	time
800 km	1 hour
?	0.2 hours

5. The earth travels around the sun at a speed of 30 km/s (kilometres per second). How many seconds would the earth take to travel 420 kilometres?
Don't include the unit in your answer.

■ 14

distance	time
30 km	1 second
420 km	?

6. A train is travelling at 120 km/h (kilometres per hour). How many hours does it take to travel 600 kilometres?
Don't include the unit in your answer.

■ 5

distance	time
120 km	1 hour
600 km	?

7. A hovercraft is travelling at 48 km/h (kilometres per hour). How many **minutes** does it take to travel 16 kilometres?
Don't include the unit in your answer.

■ 20

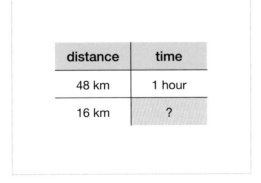

distance	time
48 km	1 hour
16 km	?

8. A cheetah can run 30 metres in a second. How long does it take to run 75 metres?
Give your answer as a decimal and don't include the units.

■ 2.5

distance	time
30 m	1 second
75 m	?

9. A car is travelling at 120 km/h (kilometres per hour). How many kilometres does it travel in quarter of an hour?
Don't include the unit in your answer.

■ 30

distance	time
120 km	1 hour
?	$\frac{1}{4}$ hour

10. A cyclist is riding at 30 km/h (kilometres per hour). How many kilometres can she ride in 0.3 hours?
Don't include the unit in your answer.

■ 9

distance	time
30 km	1 hour
?	0.3 hours

Level 2: Fluency - Calculating speed from written information and distance time graphs.

✹ **Required:** 7/10 ✹ **Student Navigation:** on
✹ **Randomised:** off

11. A car travels 300 kilometres in 4 hours. What is the average speed of the car in kilometres per hour?
Don't include the unit in your answer.

▪ 75

	distance	time
	300 km	4 hours
average speed	? km/h	1 hour

12. A school group completes a 21 kilometre walk in 6 hours. What is the average speed of the group in kilometres per hour?
Give your answer as a decimal and don't include the units.

▪ 3.5

	distance	time
	21 km	6 hours
average speed	? km/h	1 hour

13. A motorcyclist travels 42 kilometres in half an hour. What is the average speed of the motorcyclist in kilometres per hour?
Don't include the unit in your answer.

▪ 84

	distance	time
	42 km	$\frac{1}{2}$ hour
average speed	? km/h	1 hour

14. The graph shows the distance a lorry travels on a journey. Calculate the average speed of the lorry in kilometres per hour.
Don't include the unit in your answer.

▪ 80

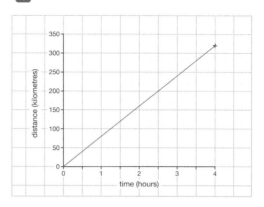

15. An athlete can run 400 metres in one minute. What speed is this in kilometres per hour?
Don't include the unit in your answer.

▪ 24

	distance	time
	400 m	1 minute
speed	? km/h	1 hour

16. Going to school, Grace walks part of the way and then travels by bus for the remainder of the journey. Calculate Grace's average speed for the total journey in kilometres per hour.
Don't include the unit in your answer.

▪ 16

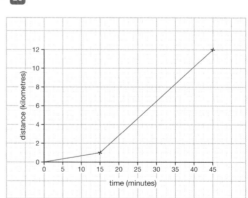

17. A shark can swim 10 metres per second. What is
a its speed in kilometres per hour?
b *Don't include the unit in your answer.*
c

■ 36

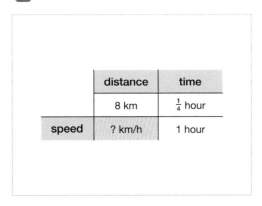

18. A cyclist travels 8 kilometres in quarter of an hour.
1 What speed is the cyclist travelling at in kilometres
2 per hour?
3 *Don't include the unit in your answer.*

■ 32

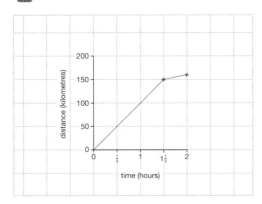

19. The graph shows Cindy's journey from
1 Birmingham to London.
2 What is Cindy's average speed for the total
3 journey in kilometres per hour?
Don't include the unit in your answer.

■ 80

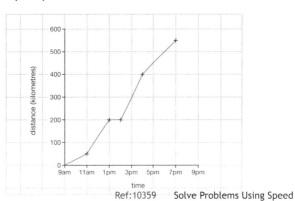

20. A train travels 50 metres in one second. What
a speed is this in kilometres per hour?
b *Don't include the unit in your answer.*
c

■ 180

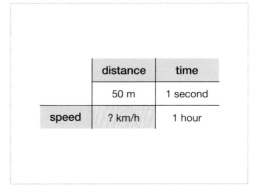

Level 3:   Reasoning - Using the speed formula in
different contexts.

✱ **Required:** 5/5   ✱ **Student Navigation:** on
✱ **Randomised:** off

21. If speed = distance travelled ÷ time taken, what is
the correct formula to calculate time taken?

1/4   ■ speed ÷ distance travelled ■ speed × distance travelled
■ distance travelled ÷ speed   ■ distance travelled × speed

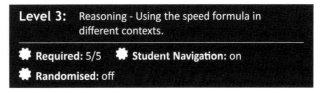

$$speed = \frac{distance\ travelled}{time\ taken}$$

22. Usain Bolt can run 100 metres in less than 10
a seconds. Carl says that Usain can run 1 kilometre
b in less than 100 seconds. Is Carl correct? Explain
c your answer.

23. The graph shows Mary's journey from London to
Glasgow. Between which hours was Mary
travelling the fastest?

1/5   ■ 9am - 11am ■ 11am - 1pm ■ 1pm - 2pm ■ 2pm - 4pm
■ 4pm - 7pm

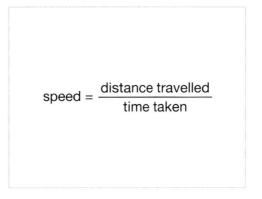

24. An ostrich has a top speed of 72 km/h. What is this speed measured in metres per second?
Don't include the unit in your answer.

■ 20

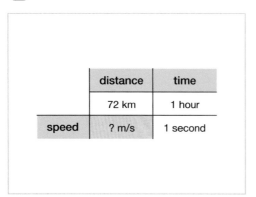

	distance	time
	72 km	1 hour
speed	? m/s	1 second

25. Callie and Sophia are 20 kilometres apart and set off walking towards each other at the same time. Callie travels at 3 km/h and Sophia travels at 5 km/h.
How many hours will it take them to meet up?
Don't include the units and give your answer as a decimal.

■ 2.5

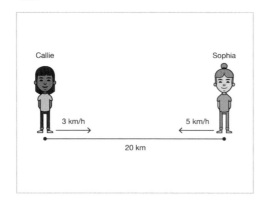

26. Ralph travels 60 kilometres in 40 minutes, Suleman travels 30 kilometres in 24 minutes. How much faster is Ralph travelling than Suleman?
Give your answer in km/h but don't include the units.

■ 15

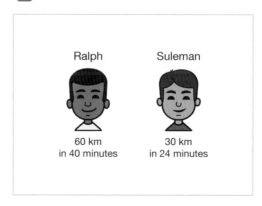

Ralph   Suleman

60 km   30 km
in 40 minutes   in 24 minutes

27. Two cars go on a journey starting from the same place. The first car travels at 90 km/h, the second sets off half an hour later and travels at 100 km/h. How many hours does it take the second car to catch up with the first?

■ 4.5

28. Charlie drives 10 kilometres at 30 km/h and 10 kilometres at 60 km/h. What is Charlie's average speed in kilometres per hour?
Don't include the unit in your answer.

■ 40

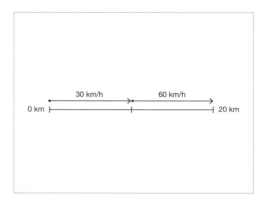

30 km/h   60 km/h
0 km |————————|————————| 20 km

29. Lewis cycles at 25 km/h and takes 30 minutes to reach his friends' house. On the way home he cycles at the same speed but stops to repair a puncture. His average speed is now 20 km/h because of the stop.
How many minutes does Lewis stop for to repair the puncture?
Give your answer as a decimal and don't include the unit.

■ 7.5

30. The tortoise and the hare are having a 2 kilometre race.

The tortoise runs the whole race at a constant speed of 5 km/h. The hare runs at 20 km/h for half the race and then has a sleep for 20 minutes.
What speed in km/h does the hare have to run the second half of the race to catch up with the tortoise on the finish line?
Don't include the unit in your answer.

▪ 60

Mathematics

Geometry

3D Shape

Use the properties of 3D shapes to solve problems

Competency: Use the properties of faces, surfaces, edges and vertices of cubes, cuboids, prisms, cylinders, pyramids, cones and spheres to solve problems in 3D.

Quick Search Ref: 10094

Correct: Correct.   **Wrong:** Incorrect, try again.   **Open:** Thank you.

Level 1: Use the properties of 3D shapes to solve problems.

⚙ **Required:** 10/10   ⚙ **Student Navigation:** on   ⚙ **Randomised:** off

1. Which of the nets folds to form a cube?
There are 5 correct answers.

▢
☒
▢
5/6

▪ (i) ▪ (ii) ▪ (iii) ▪ (iv) ▪ (v) ▪ (vi)

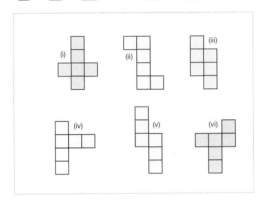

2. Jay makes a shape by joining two square-based pyramids at their bases. How many vertices does the new shape have?

▪ 6

3. If a solid square-based pyramid has **all** of its vertices cut off as shown in the diagram, how many edges will the new shape have?

▪ 24

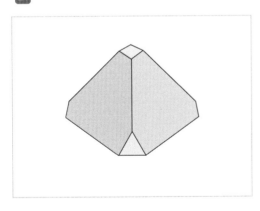

4. How many identical triangles are needed to make this icosahedron shown?

▪ 20

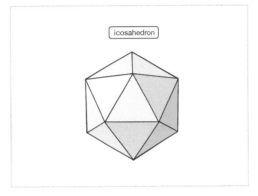

icosahedron

5. The image shows two **similar** cuboids. What is the volume of the larger cuboid?

▪ 23,328 ▪ 23328

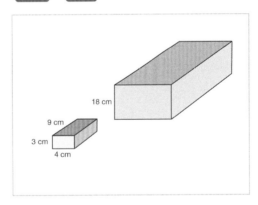

18 cm

9 cm

3 cm

4 cm

6. A shape has eight faces and six vertices. How many edges does it have?

▪ 12

7. What edge will *A* meet when the net is folded to make a 3D shape?

■ B ■ C ■ D ■ E ■ F ■ G ■ H

1/7

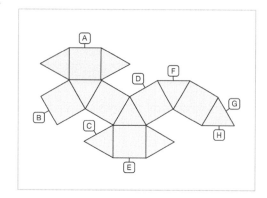

8. How many planes of symmetry does a cube have?

■ 9

9. Place the shapes in ascending order (smallest first) according to the number of edges they have.

■ Sphere ■ Cone ■ Tetrahedron ■ Trapezium prism
■ Dodecahedron

10. A platonic solid is a convex, 3D shape with faces that are all identical. How many platonic solids are there? Explain your answer.

Mathematics

Statistics

Pie Charts

Averages and Range

Stem and Leaf

Interpret Pie Charts and Calculate Angles

Competency: Construct and interpret appropriate tables, charts, and diagrams, including frequency tables, bar charts, pie charts, and pictograms for categorical data, and vertical line (or bar) charts for ungrouped and grouped numerical data.

Quick Search Ref: 10425

Correct: Correct.   **Wrong:** Incorrect, try again.   **Open:** Thank you.

Level 1:   Understanding – Interpreting pie charts.

✱ **Required:** 7/10   ✱ **Student Navigation:** on   ✱ **Randomised:** off

1. The unit of measurement for angles is:

 ▪ degrees

2. How many degrees are there in a circle?
Don't include the ° (degrees) symbol in your answer.

▪ 360

3. Which pie chart correctly represents the information in the table?

 ▪ (i) ▪ (ii) ▪ (iii)

1/3

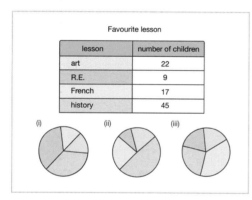

Favourite lesson

lesson	number of children
art	22
R.E.	9
French	17
history	45

(i)   (ii)   (iii)

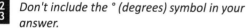

4. What angle would represent strawberry ice cream in a pie chart?
Don't include the ° (degrees) symbol in your answer.

▪ 65

Favourite flavours of ice cream

flavour	number of children	angle in pie chart
chocolate	18	90°
strawberry	13	?
mint	9	45°
vanilla	28	140°
pistachio	4	20°

5. If Ally spends £30 on vegetables at the supermarket, how much does she spend in total?
Include the £ sign in your answer.

▪ £120.00 ▪ £120

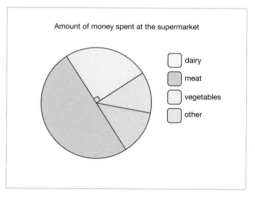

Amount of money spent at the supermarket

□ dairy
▦ meat
□ vegetables
□ other

6. What is the angle of the sector that represents cats?
Don't include the ° (degrees) symbol in your answer.

▪ 180

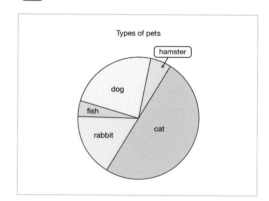

Types of pets

hamster

dog

fish

rabbit   cat

7. 25% of the students chose green as their favourite colour, what angle will represent this?
Don't include the ° (degrees) symbol in your answer.

a
b
c

▪ 90

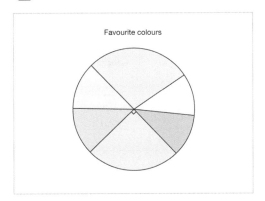

8. The total of £1 coins and 50p coins in Reba's money box is £12. How much does she have in her money box altogether?
Include the £ sign in your answer.

a
b
c

▪ £24.00 ▪ £24

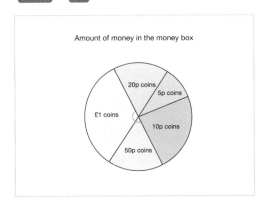

9. What is the angle of the sector that represents the girls?
Don't include the ° (degrees) symbol in your answer.

a
b
c

▪ 270

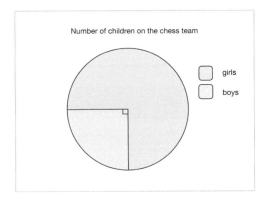

10. What angle would represent rugby in a pie chart?
Don't include the ° (degrees) symbol in your answer.

1
2
3

▪ 93

Favourite sport of Year 8 pupils

sport	netball	tennis	rugby	cross-country
number of children	45	28	31	16
angle in pie chart	135°	84°	?°	48°

Level 2: Fluency – Calculating angles to construct a pie chart.

✱ **Required:** 7/10   ✱ **Student Navigation:** on
✱ **Randomised:** off

11. 72 people take part in a survey. Calculate the angle that represents one person in a pie chart.
Don't include the ° (degrees) symbol in your answer.

1
2
3

▪ 5

12. In a pie chart, what angle would represent the number of games lost by the football team?
Don't include the ° (degrees) symbol in your answer.

1
2
3

▪ 72

Football team results

result	number of games	fraction of total	angle in pie chart
won	7	$\frac{7}{20}$	126°
drew	1		
lost	4		?°
total	20	1	360°

Level 2: *cont.*

13. Calculate the angle in a pie chart that would represent the pupils who speak Urdu as their first language.

Don't include the ° (degrees) symbol in your answer.

▪ 35

Pupils' first language

language	number of pupils	fraction of total	angle in pie chart
English	96		
Spanish	4		
Polish	25		
Urdu	14		?°
Mandarin	5		
total		1	360°

14. Calculate the angle in a pie chart that would represent the number of science teachers.
a b c
Don't include the ° (degrees) symbol in your answer.

▪ 96

Teaching subjects

subject	maths	English	science	total
number of teachers	ЖЖ ЖЖ ЖЖ lll	ЖЖ ЖЖ ЖЖ	ЖЖ ЖЖ ll	
fraction of total				
angle in pie chart				

15. Calculate the angle in a pie chart that would represent computer games?
1 2 3
Don't include the ° (degrees) symbol in your answer.

▪ 100

Favourite game type

game type	fraction	angle in pie chart
computer		?°
console	$\frac{1}{3}$	
board	$\frac{1}{9}$	
outdoor	$\frac{5}{18}$	
total	1	360°

16. 60 people are asked their favourite flavour of crisps. 12 said plain, 31 said salt and vinegar, 8 said cheese and onion and 9 said prawn cocktail. Calculate the angle of the sector that would represent salt and vinegar in a pie chart.
a b c
Don't include the ° (degrees) symbol in your answer.

▪ 186

Favourite flavour of crisps

flavour	fraction of total	angle in pie chart
plain		
salt and vinegar		
cheese and onion		
prawn cocktail		
total	1	360°

17. What angle in a pie chart would represent the population of Africa?
a b c
Give your answer to 1 decimal place and don't include the ° (degrees) symbol.

▪ 61.2

Approximate world population by continent

continent	% (approx) of world population	fraction of population	angle in pie chart
Asia	59%		
Africa	17%		?°
Europe	10%		
North America	8%		
South America	5%		
Oceania	1%		
Antartica	0%		
total		1	360°

18. Calculate the angle in a pie chart that would represent the children who travel by bus to school.
1 2 3
Don't include the ° (degrees) symbol in your answer.

▪ 174

How children travel to school

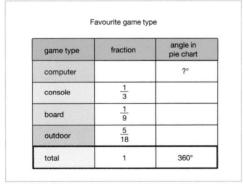

transport	walk	bus	car	train	total
number of children	48	97	36	9	
fraction of total					1
angle in pie chart		?°			360°

19. What angle in a pie chart would represent the number of Year 11s in drama club?
a b c
Don't include the ° (degrees) symbol in your answer.

▪ 126

Year groups of students in drama club

year group	fraction/ percentage of total	angle in pie chart
7	$\frac{3}{20}$	
8	$\frac{1}{5}$	
9	5%	
10	$\frac{1}{4}$	
11	35%	
total	1	360°

20. 36 people were asked their favourite brand of car. 4 people said Fiat, 12 people said Vauxhall, 15 people said Honda and the rest said Seat. Calculate the angle in a pie chart that would represent Seat.
a b c
Don't include the ° (degrees) symbol in your answer.

▪ 50

Favourite brand of car

brand	number of people	fraction of total	angle in pie chart
Fiat			
Vauxhall			
Honda			
Seat			
total		1	360°

21. How many children chose orange as their favourite juice?
a b c
▪ 24

Favourite fruit juice

juice	number of people	angle in pie chart
apple	12	90°
orange	?	180°
pineapple	5	
guava		52.5°
cranberry	0	0°
total		360°

22. Year 9 pupils were asked if they like their uniform. What does the pie chart tell you?
a b c

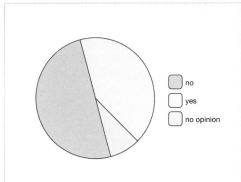

no
yes
no opinion

23. Your teacher has asked you to survey your peers about their favourite sports and display the results in a pie chart. Explain how you would decide how many people to ask.
a b c

24. Select the statements that are true.

☐ ☒ ☐
2/5

▪ **There are more pupils in Year 7 than any other year group.**
▪ **There are more Year 9 boys than Year 11 girls.**
▪ **¾ of the pupils are not in Year 7.**
▪ **There are 90 children in Year 7.**
▪ **There are the same number of pupils in Year 10 as in Year 11.**

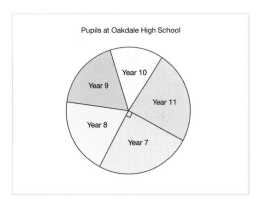

Pupils at Oakdale High School

25. Stevie says, "More boys go to Ashton High than Redbrick Prep". Is Stevie correct? Explain your answer.
a b c

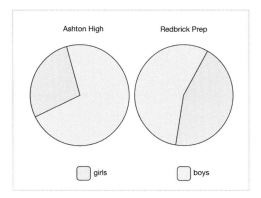

Ashton High   Redbrick Prep

girls   boys

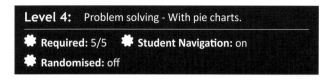

Level 4:   Problem solving - With pie charts.

❖ **Required:** 5/5   ❖ **Student Navigation:** on
❖ **Randomised:** off

26. A total of £588 was raised at the school bake sale.
a The cake stall raised £217.56 and the drinks stall
b raised £105.84. What percentage of the money
c was raised by the brownie stall?
Include the % (percent) sign in your answer.

▪ 20%

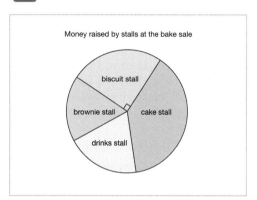

27. Oscar has a bag of red, green and yellow marbles
a in the ratio 1:3:4. Calculate the angle of the green
b segment if you were to record this in a pie chart.
c *Don't include the ° (degrees) symbol in your*
answer.

▪ 135

28. In a survey 48 people said they are going on
a holiday to Spain. How many more people are
b going on holiday to England, America or Australia
c than are going on holiday to Scotland?

▪ 90

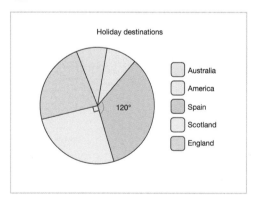

29. Jodie asked everyone in Year 8 what their favourite
a film genre was, so she could record the
b information in a pie chart. How many people
c chose horror?

▪ 32

Favourite film genre		
genre	number of people	angle in pie chart
action	38	57°
comedy	57	
romance	5	73.5°
horror		
sci-fi	43	0°
fantasy		31.5°
total		

30. 45% of students prefer maths. 40 more students
a prefer maths to English. How many students prefer
b science?
c

▪ 60

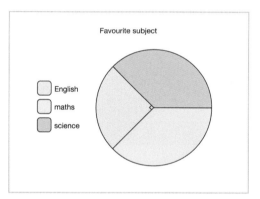

Calculate the mean, median, mode, range and outlier

Competency: Describe, interpret and compare observed distributions of a single variable through: appropriate graphical representation and appropriate measures of central tendency (mean, mode, median) and spread.

Quick Search Ref: 10150

Correct: Correct.   Wrong: Incorrect, try again.   Open: Thank you.

Level 1: Understanding - Calculating the mean, median, mode and range of a data set.

✿ **Required:** 7/10   ✿ **Student Navigation:** on   ✿ **Randomised:** off

---

1. What is the **mean** value of the following data set?
3, 4, 4, 5, 6, 7, 8, 8, 8, 9.

▪ **6.2**

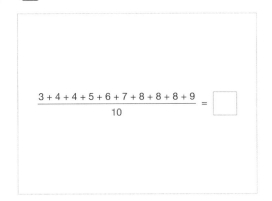

$$\frac{3+4+4+5+6+7+8+8+8+9}{10} = \boxed{}$$

2. Find the **mode** of the following data set:
3, 4, 4, 5, 6, 7, 8, 8, 8, 9.

▪ **8**

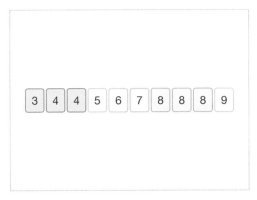

3 4 4 5 6 7 8 8 8 9

3. Calculate the **range** of the following data set:
3, 4, 4, 5, 6, 7, 8, 8, 8, 9.

▪ **6**

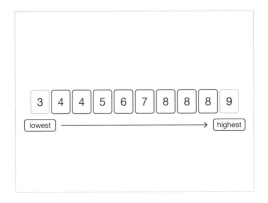

3 4 4 5 6 7 8 8 8 9
lowest ――――――――→ highest

4. What is the **median** value of the following data set?
3, 3, 4, 4, 6, 7, 7, 8, 8, 8, 8.

▪ **7**

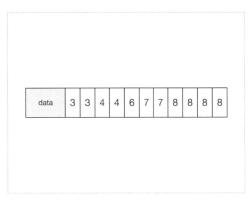

| data | 3 | 3 | 4 | 4 | 6 | 7 | 7 | 8 | 8 | 8 | 8 |

5. Eight children in class 6 record their height in a table. Select the two values for the **mode** height.

▪ 127 ▪ 132 ▪ **133** ▪ 134 ▪ 135 ▪ **137**

2/6

name	height (cm)
Danny	137
Luke	132
Nathan	127
Louise	133
Helen	135
Isaac	137
Maya	133
Joseph	134

6. Which value is the **outlier** in the following data set?
6, 86, 92, 94, 96, 96.

▪ **6**

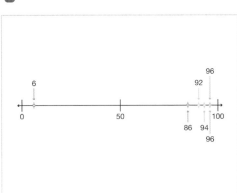

7. There are 10 players in a netball squad. The coach records the number of games each girl plays in the season. What is the **median** number of games played by each girl?

1/4   18, 18, 14, 15, 14, 17, 13, 15, 18, 17.

▪ **15** ▪ **16** ▪ **17** ▪ **18**

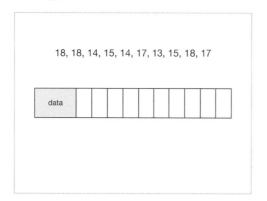

8. Mr Farran's class have a quiz and each group gets a score out of 50. What is the **range** of the results?

▪ **23**

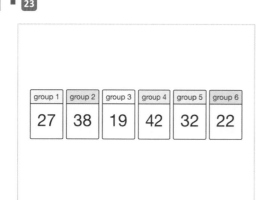

9. Find the **mode** of the following data set:
12, 8, 9, 12, 15, 13, 15, 19, 15, 5.

▪ **15**

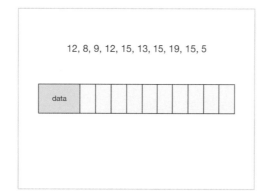

10. What is the **median** of the following data set?
15, 8, 9, 12, 13, 15, 16, 15, 5.

▪ **13**

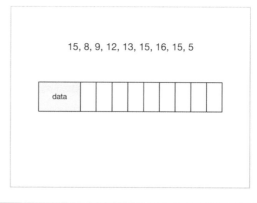

Level 2:   Fluency - Finding the mean, median, mode and range in context (including charts).

🌸 **Required:** 7/10   🌸 **Student Navigation:** on
🌸 **Randomised:** off

11. Calculate the mean, median and mode of the following data set and select the one with the **highest** value:

1/3   8, 15, 6, 8, 9, 9, 8, 12, 10, 11.

▪ **mean** ▪ **median** ▪ **mode**

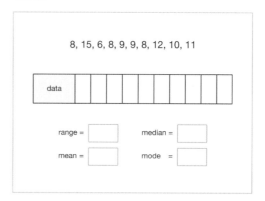

12. A shop has three bags of apples left at the end of the day. From these remaining bags, calculate the **mean** weight of an apple.
Include the unit g (grams) in your answer.

▪ **25 grams** ▪ **25 g**

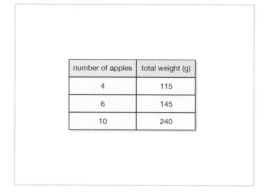

Level 2: cont.

13. The bar graph shows the shoe size of each person who goes to a bowling alley for a birthday party. What is the **mode** shoe size?

   ▪ 6

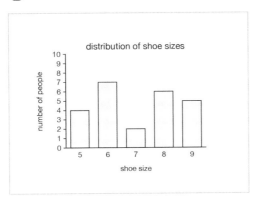

14. Gavin asked his friends how many books they bought last year. He can't remember one of their answers, but he knows the median is 7.5 books. What is the missing number?

   ▪ 8

15. Barry is playing darts and puts his first five scores into a bar chart. What is the **range** of his scores?

   ▪ 28

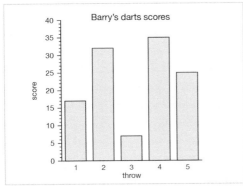

16. Franky wants to know what flavour of ice-cream is the most popular. What is the **mode** flavour of ice-cream?

1/5 ▪ strawberry ▪ vanilla ▪ chocolate ▪ mint ▪ raisin

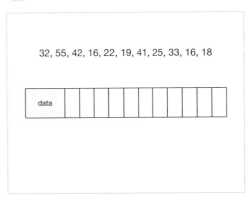

flavour	ice-creams sold
strawberry	13
vanilla	8
chocolate	17
mint	14
raisin	8

17. The chart shows the ages of the people who go to the cinema. What is the age **range**?

   ▪ 39

32, 55, 42, 16, 22, 19, 41, 25, 33, 16, 18

data										

18. A doctor's surgery records the age of its patients. What is the **median** age?

   ▪ 45

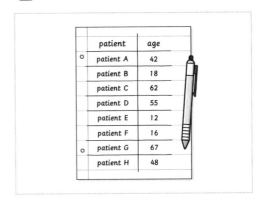

patient	age
patient A	42
patient B	18
patient C	62
patient D	55
patient E	12
patient F	16
patient G	67
patient H	48

Level 2: *cont.*

19. Jasmin buys a new book each week in the summer holidays. What is the **mean** cost of the books she buys?
Include the £ sign in your answer.

a
b
c

▪ £7.00 ▪ £7

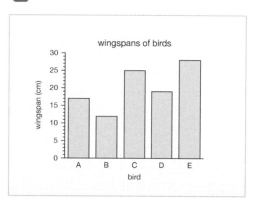

week	cost of book (£)
1	8.00
2	7.00
3	5.00
4	9.00
5	6.00
6	7.00

20. Wesley measures the wingspan of all the birds that land in his garden in an afternoon. What is the **range** of the wingspans?
Do not include the units in your answer.

1
2
3

▪ 16

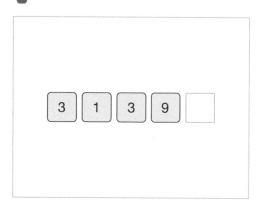

Level 3:   Reasoning - Negative numbers, missing numbers and outliers.

❋ **Required:** 5/5   ❋ **Student Navigation:** on
❋ **Randomised:** off

21. A group of five numbers have a mean of 4, a median of 3 and a mode of 3. What number is missing from the group?

1
2
3

▪ 4

3 1 3 9 ☐

22. The **range** of the following data set is 44 and the missing number has 3 digits. What is the missing number?
72, 72, 66, 82, 68, ?

1
2
3

▪ 110

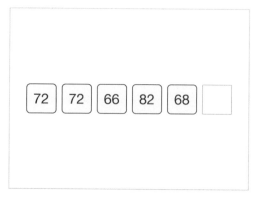

72 72 66 82 68 ☐

23. Alex's teacher calculates the mean score for a maths test. What can Alex's teacher do to calculate a more accurate average of the scores?

a
b
c

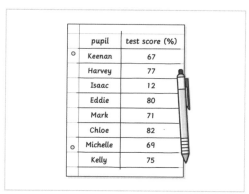

pupil	test score (%)
Keenan	67
Harvey	77
Isaac	12
Eddie	80
Mark	71
Chloe	82
Michelle	69
Kelly	75

24. Grace says, "To find the mean of the following numbers, I can round each number to 100, add them together and divide by 8".
Will Grace's answer be accurate? Explain your answer.
105, 103, 112, 96, 101, 106, 99, 110.

a
b
c

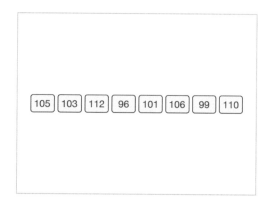

105 103 112 96 101 106 99 110

Level 3: *cont.*

25. Danny and Eddy are supposed to have six number cards each but Danny has taken one too many. Which card can Danny give to Eddy so that they both have the same **range** of numbers?

▪ **7**

Level 4: Problem Solving - Calculating missing values using mean, median, mode and range.

❋ **Required:** 5/5   ❋ **Student Navigation:** on
❋ **Randomised:** off

26. Katie and her four friends each think of a number between 1 and 9. The sequence of the five numbers has the following rules:
The mean, median, mode and range = 5;
Only one appears twice;
The lowest number chosen is 2.
Write the five numbers chosen by Katie and her friends as a number sequence, in the format, 1, 2, 3, 4, 5.

▪ **2, 5, 5, 6, 7.** ▪ **2, 5, 5, 6, 7** ▪ **2, 5, 5, 6, 7,**

27. There are 46 football matches in one week. The number of goals scored in each match is recorded in a line graph. What is the **median** number of goals scored in a single match?

▪ **2.5**

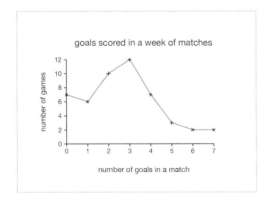

28. Amy lives with her Mum, Dad and three brothers. Each member of the family has a job and their mean earnings are £12,000 per year. When Amy moves out, the mean earnings of the household increase to £12,400. How much does Amy earn?
Include the £ sign in your answer.

▪ **£10,000** ▪ **£10000**

29. The value of Y on the number line is 21. What is the **range** of X, Y and Z?

▪ **77**

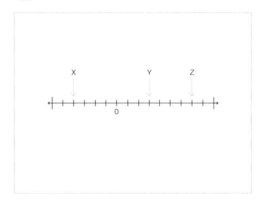

30. John enters an 8 mile run. He runs 2 miles at 6 miles per hour (mph) and 1 mile at 4 miles per hour.
On average, how many minutes does John need to complete each remaining mile to finish in 75 minutes?
Don't include units in your answer.

▪ **8**

Draw and Interpret Stem and Leaf Diagrams

Competency: Draw and interpret appropriate tables, charts, and diagrams, including frequency tables, bar charts, pie charts, and pictograms for categorical data, and vertical line (or bar) charts for ungrouped and grouped numerical data.

Quick Search Ref: 10471

Correct: Correct.   Wrong: Incorrect, try again.   Open: Thank you.

Level 1: Understanding - Reading and drawing stem and leaf diagrams.

✿ **Required:** 7/10   ✿ **Student Navigation:** on   ✿ **Randomised:** off

---

1. Katie is making a stem and leaf diagram of the ages of people who work in her office. What digit should she fill in as the next leaf on her diagram?

▪ 6

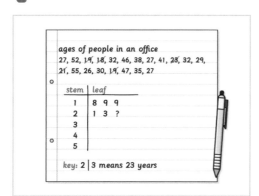

2. Here is a stem and leaf diagram of the ages of people who work in an office. What age is the youngest person in the office?

▪ 18

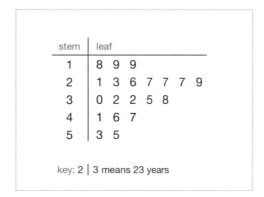

3. Gulab is making a stem and leaf diagram of the heights of students in his class. What is the stem part of the next value he fills in?

▪ 14

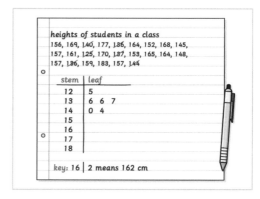

4. Looking at the stem and leaf diagram, what height is the tallest person in the class?
Include the units cm (centimetres) in your answer.

▪ 183 centimetres   ▪ 183 cm

stem	leaf
12	5
13	6 6 7
14	0 4 5 8
15	2 3 6 7 7 7 9
16	1 4 4 5 8 9
17	0 7
18	3

key: 15 | 3 means 153 years

5. Here is a stem and leaf diagram of the ages of people who work in an office. How many people work in the office?

■ 20

stem	leaf
1	8 9 9
2	1 3 6 7 7 7 9
3	0 2 2 5 8
4	1 6 7
5	3 5

key: 2 | 3 means 23 years

6. A nurse records the temperature of patients on her ward in a stem and leaf diagram. What digit should they fill in as the next leaf on their diagram?

■ 2

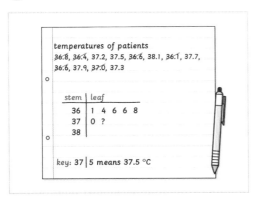

temperatures of patients
36.8, 36.4, 37.2, 37.5, 36.6, 38.1, 36.1, 37.7, 36.6, 37.9, 37.0, 37.3

stem	leaf
36	1 4 6 6 8
37	0 ?
38	

key: 37 | 5 means 37.5 °C

7. A nurse records the temperature of patients on her ward in a stem and leaf diagram. What is the highest temperature of all the patients?
Don't include the units in your answer.

■ 38.1

stem	leaf
36	1 4 6 6 8
37	0 2 3 5 7 9
38	1

key: 37 | 5 means 37.5 °C

8. Here is a stem and leaf diagram of the ages of people who work in an office. What age is the oldest person in the office?

■ 55

stem	leaf
1	8 9 9
2	1 3 6 7 7 7 9
3	0 2 2 5 8
4	1 6 7
5	3 5

key: 2 | 3 means 23 years

9. Here is a stem and leaf diagram of the heights of students in a class in centimetres. How many students are in the class?

■ 24

stem	leaf
12	5
13	6 6 7
14	0 4 5 8
15	2 3 6 7 7 7 9
16	1 4 4 5 8 9
17	0 7
18	3

key: 15 | 3 means 153 years

10. A nurse records the temperature of patients on her ward in a stem and leaf diagram. What is the lowest temperature of all the patients?
Don't include the units in your answer.

■ 36.1

stem	leaf
36	1 4 6 6 8
37	0 2 3 5 7 9
38	1

key: 37 | 5 means 37.5 °C

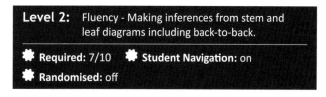

Level 2: Fluency - Making inferences from stem and leaf diagrams including back-to-back.

✱ **Required:** 7/10   ✱ **Student Navigation:** on
✱ **Randomised:** off

11. Here is a stem and leaf diagram of the ages of people who work in an office. What is the **mode** of their ages?

▪ 27

```
1 | 8 9 9
2 | 1 3 6 7 7 7 9
3 | 0 2 2 5 8
4 | 1 6 7
5 | 3 5

key: 2 | 3 means 23 years
```

12. A nurse records the temperature of patients on her ward in a stem and leaf diagram. What is the median temperature of the patients?
Don't include the units in your answer.

▪ 37.1

```
36 | 1 4 6 6 8
37 | 0 2 3 5 7 9
38 | 1

key: 37 | 5 means 37.5 °C
```

13. A biologist makes a stem and leaf diagram of the weight of salmon in a river. Which weight have they missed from the diagram?
Include the units kg (kilograms) in your answer.

▪ 2.2 kilograms   ▪ 2.2 kg

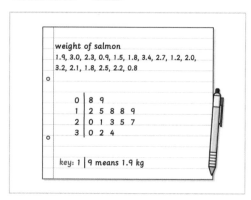

14. Petra is making a back-to-back stem and leaf diagram of the times boys and girls took to run 400 metres. What is the next digit she should fill in on her diagram?

▪ 0

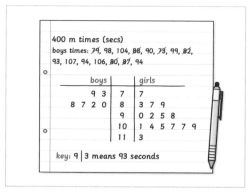

15. A biologist makes a stem and leaf diagram of the weights of salmon in a river. What is the **range** of the weights of the salmon?
Include the units kg (kilograms) in your answer.

▪ 2.6 kg   ▪ 2.6 kilograms

```
0 | 8 9
1 | 2 5 8 8 9
2 | 0 1 2 3 5 7
3 | 0 2 4

key: 2 | 4 means 2.4 kg
```

16. The back-to-back stem and leaf diagram shows the times boys and girls took to run 400 metres. What is the fastest time of the boys?
Give your answer in seconds and don't include the units.

▪ 73

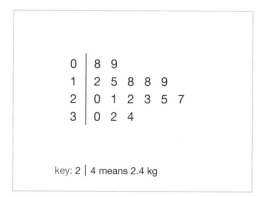

Level 2: *cont.*

17. The back-to-back stem and leaf diagram shows boys and girls times to run 400 metres. What is the **range** of the times for the boys?
Give your answer in seconds and don't include the units.

- 34

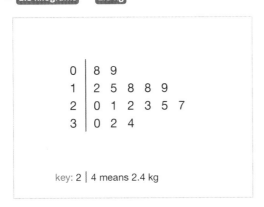

key: 8 | 4 means 84 seconds

18. A biologist makes a stem and leaf diagram of the weight of salmon in a river. What is the **mode** weight of the salmon?
Include the units kg (kilograms) in your answer.

- 1.8 kilograms ▪ 1.8 kg

```
0 | 8 9
1 | 2 5 8 8 9
2 | 0 1 2 3 5 7
3 | 0 2 4
```

key: 2 | 4 means 2.4 kg

19. Here is a stem and leaf diagram of the heights of students in a class. What is the **range** of their heights?
Include the units cm (centimetres) in your answer.

- 58 cm ▪ 58 centimetres

```
12 | 5
13 | 6 6 7
14 | 0 4 5 8
15 | 2 3 6 7 7 7 9
16 | 1 4 4 5 8 9
17 | 0 7
18 | 3
```

key: 15 | 3 means 153 years

20. The back-to-back stem and leaf diagram shows the times boys and girls took to run 400 metres. What is the **median** of the times for the boys?
Give your answer in seconds and don't include the units.

- 93

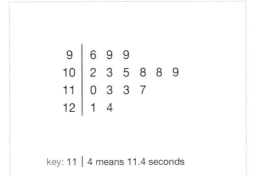

key: 8 | 4 means 84 seconds

Level 3: Reasoning - Misconceptions and comparing data in stem and leaf diagrams.

✿ **Required:** 5/5   ✿ **Student Navigation:** on
✿ **Randomised:** off

21. Joanne is reading a stem and leaf diagram of the times people take to complete a puzzle. Joanne says the slowest time is 124 seconds. Is Joanne correct? Explain your answer.

```
 9 | 6 9 9
10 | 2 3 5 8 8 9
11 | 0 3 3 7
12 | 1 4
```

key: 11 | 4 means 11.4 seconds

22. The back-to-back stem and leaf diagram shows the times boys and girls took to run 400 metres. Select the statements which correctly compare their times.

2/4

- ■ The mode time is slower for the boys than for the girls.
- ■ The range of times is smaller for the boys than for the girls.
- ■ The range of times is larger for the boys than for the girls.
- ■ The mode time is faster for the boys than for the girls.

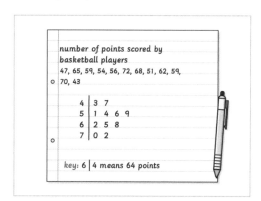

boys		girls
9 3	7	7
8 7 2 0	8	3 7 9
9 8 4 4 3 0	9	0 2 5 8
7 6 4	10	1 4 5 7 7 9
	11	3

key: 8 | 4 means 84 seconds

23. Ratib is making a stem and leaf diagram of the number of points scored by his teammates in a basketball tournament. What mistake has Ratib made? Explain your answer.

a
b
c

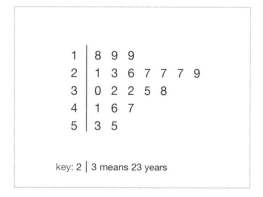

number of points scored by
basketball players
47, 65, 59, 54, 56, 72, 68, 51, 62, 59,
o 70, 43

4	3 7
5	1 4 6 9
6	2 5 8
7	0 2

o

key: 6 | 4 means 64 points

24. Here is a stem and leaf diagram of the age of people who work in an office. What is the modal class interval?

1/5   ■ 10–19 ■ 20–29 ■ 30–39 ■ 40–49 ■ 50–59

1	8 9 9
2	1 3 6 7 7 7 9
3	0 2 2 5 8
4	1 6 7
5	3 5

key: 2 | 3 means 23 years

25. Ratib records the points scored by his teammates in a basketball tournament. If Ratib now adds his own score of 78 points to the table, which of the following statements are true?

2/6

- ■ The range decreases.   ■ The range stays the same.
- ■ The range increases.   ■ The median decreases.
- ■ The median stays the same.   ■ The median increases.

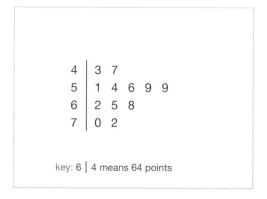

4	3 7
5	1 4 6 9 9
6	2 5 8
7	0 2

key: 6 | 4 means 64 points

✱ **Required:** 5/5   ✱ **Student Navigation:** on
✱ **Randomised:** off

26. Here is a stem and leaf diagram of the heights of students in a class. If a student is selected at random, what is the probability they are taller than 160 cm?
Give your answer as a fraction in its simplest form.

a
b
c

- ■ 3/8

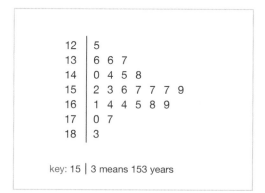

12	5
13	6 6 7
14	0 4 5 8
15	2 3 6 7 7 7 9
16	1 4 4 5 8 9
17	0 7
18	3

key: 15 | 3 means 153 years

Level 4: *cont.*

27. Here is a stem and leaf diagram of the ages of people who work in an office. The two oldest people leave the office. How many years does the mean age of people decrease by?

123

■ 6

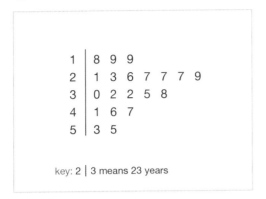

```
1 | 8  9  9
2 | 1  3  6  7  7  7  9
3 | 0  2  2  5  8
4 | 1  6  7
5 | 3  5
```

key: 2 | 3 means 23 years

28. A biologist makes a stem and leaf diagram of the weights of salmon in a river. The biologist can only carry 25 kg of fish. What is the greatest number of fish he can carry?

123

■ 13

```
0 | 8  9
1 | 2  5  8  8  9
2 | 0  1  2  3  5  7
3 | 0  2  4
```

key: 2 | 4 means 2.4 kg

29. A nurse records the temperatures of patients on her ward in a stem and leaf diagram. What is the mean temperature of the patients in degrees centigrade?
Don't include the unit in your answer.

123

■ 37.1

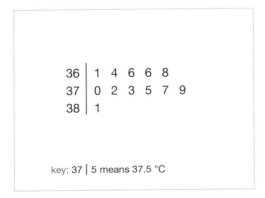

```
36 | 1  4  6  6  8
37 | 0  2  3  5  7  9
38 | 1
```

key: 37 | 5 means 37.5 °C

30. The back-to-back stem and leaf diagram shows the times boys and girls took to run 400 metres. The four fastest boys then run a 4 × 400 metre relay race against the four fastest girls. If the boys and girls run the same times as in the diagram, by how many seconds would the boys finish ahead of the girls?

a
b
c

■ 22

```
                    boys  |    | girls
                   9  3   | 7  | 7
              8  7  2  0  | 8  | 3  7  9
        9  8  4  4  3  0  | 9  | 0  2  5  8
                 7  6  4  | 10 | 1  4  5  7  7  9
                         | 11 | 3
```

key: 8 | 4 means 84 seconds

LbQ Super Deal
Class set of tablets and charging cabinet

Class charging & storage cabinet

+ 32 x
Pupil 7" HD tablets with protective cover

+ 1 x
Teacher 10" HD tablet with protective cover

*Special Offer Price

£1,100 per year on 3 years compliant operating lease

Subject to a £150 initial documentation fee

LbQ Question Set subscription required to be eligible

Min 1 LbQ subscription per set £200/year or £500/3 years

Learning by Questions App pre-loaded

Includes 3 years advanced replacement warranty on tablets (damage not covered)

Prices exclude VAT and delivery

Option to renew equipment or purchase at end of agreement

Available in United Kingdom and Republic of Ireland only

Product Code TC001

£1,100*
per year for 3 years including warranty

Place your orders with our sales partner LEB who will organise the paperwork for you:

Email: orders@lbq.org
Tel: 01254 688060

Specifications

Charging & Storage Cabinet

- 32-bay up to 10" tablet charging cabinet
- 2 easy access sliding shelves – 16 bays on each shelf
- 4 efficient quiet fans for ventilation
- Locking Doors (4 keys)
- 4 castors / 2 x handling bars Overload, leakage and lightning surge protection
- CE / ROHS / FCC compliancy
- Warranty 3 years

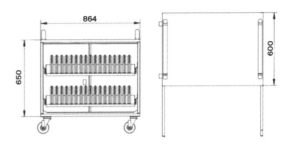

Student and Teacher tablets configuration

All tablets with LbQ Tablet Tasks App pre-loaded and installed in cabinet including charging cables and mains adapters for quick and easy deployment when onsite in classroom.

	7" Android Tablet with Protective Cover	10" Android Tablet with Protective Cover
Display		
1920 * 1200 IPS	✓	✓
16:10 Display ratio	✓	✓
Capacitive 5-touch	Capacitive 5-touch	Capacitive 10-touch
System		
Cortex 64bit Quad Core 1.5GHz CPU	✓	✓
2GB of RAM	✓	✓
16Gb of storage	✓	✓
Android 7.0	✓	✓
Front 2M and rear 5M Camera	✓	✓
Input / Output Ports		
1 x Micro SD Slot	✓	✓
1 x Micro USB (PC / device / charger)	✓	✓
Micro-HDMI output	✓	✓
1 x Earphone, 1 x Speaker, 1 x Mic	1 x Earphone, 1 x Speaker, 1 x Mic	1 x Earphone, 2 x Speaker, 1 x Mic
Communication		
Wifi – 802.11a/b/g/n	✓	✓
GPS module	✓	✓
Bluetooth	✓	✓
Power		
5V 2A	✓	✓
Battery	3000mAh battery	7000mAh battery
Physical		
Colour: Metal black	✓	✓
Weight:	230g	560g
Dimensions:	192 x 112 x 9mm (approx.)	263 x 164 x 9mm (approx.)
Warranty		
3 years Advanced Replacement for faulty tablets - does not cover damage CE / ROHS / FCC compliant	✓	✓